Business Serials of the U.S. Government

Second Edition

Business Reference and Services Section
Reference and Adult Services Division
American Library Association

Edited by **Priscilla C. Geahigan**
Robert F. Rose

American Library Association   Chicago and London 1988

Priscilla C. Geahigan is head of the reference department at the Krannert Management and Economics Library of Purdue University.

Robert F. Rose is head of the library at the Behrend College of Pennsylvania State University at Erie.

Text designed by Marcie Lange

Composed in Melior on an
 APS Micro 5 system by
 Merrill Corporation

Printed on 55-pound Glatfelter
 B-16, a pH-neutral stock,
 and bound in 10-point Carolina
 cover stock by Patterson Printing

Library of Congress Cataloging-in-Publication Data

Business serials of the U.S. Government.

    Includes index.
    1. United States—Economic conditions—Periodicals—Bibliography. 2. United States—Government publications—Bibliography. I. Geahigan, Priscilla C. II. Rose, Robert F. III. American Library Association. Business Reference and Services Section.
Z7165.U5B88   1988   016.330973   88-3428
[HC103]
ISBN 0-8389-3349-1

Copyright © 1988 by the American Library Association.

    All rights reserved except those which may be granted by Sections 107 and 108 of the Copyright Revision Act of 1976.

    Printed in the United States of America.

CONTENTS

	Preface	v
	Selected Agencies and Other Abbreviations	ix
	Abbreviations of Index Titles	x
I.	General Sources	1
II.	Economic Conditions	6
III.	Demographics	12
IV.	International Business	15
V.	Industry	22
	A. General	22
	B. Banking and Finance	24
	C. Communications	25
	D. Construction	26
	E. Manufacturing	27
	F. Natural Resources	28
	1. Mining	28
	2. Fishing	29
	3. Petroleum and Natural Gas	30
	G. Retail and Wholesale Trade	33
	H. Service	35
	1. General	35
	2. Health Care	36

	I.	Transportation	37
		1. General	37
		2. Airlines	38
		3. Highways	40
		4. Maritime	41
		5. Railroads	42
	J.	Travel	43
	K.	Utilities	44
VI.	Agriculture		48
VII.	Environment		53
VIII.	Labor		54
IX.	Small Business		59
X.	Patents and Trademarks		62
XI.	Government Grants and Contracts		64
XII.	Public Finance		66
XIII.	Taxation		70
XIV.	Consumers		75
	Title Index		78
	Subject Index		83

PREFACE

As the information explosion continues, so does the demand for more extensive current data. While generally true for most disciplines, this demand has been especially evident in the field of business. The U.S. government has traditionally been one of the best sources of business information but the information contained in its publications has not always been readily accessible to library users, especially those in non-depository libraries. Thus this book, designed to provide easy access to some of the government's most valuable publications—its business serials.

In 1978, the first edition of *Business Serials of the U.S. Government* appeared, prepared under the auspices of the Business Reference Services Committee of the Reference and Adult Services Division of ALA, and edited by Richard L. King, committee chair. Librarians across the country found it to be quite useful, although it was criticized for its lack of subject indexing. Its usefulness declined, however, as changes in the government's publication program took place—changes heightened by privatization initiatives instituted since 1981.

The Business Reference Services Committee decided in 1985 that a revised edition of this publication would be very helpful and a Publications Subcommittee was formed to undertake the revision. The subcommittee essentially started from scratch: defining our audience, the scope of the publication, the sections and titles to be included, and the format to be used, basing our list initially on those titles included in the previous edition. Readers familiar with the first edition will note that the second edition is lengthier than the first and that some titles in the first edition are not included in the second. The expanded length and the title changes reflect the committee's opinion as to the titles most useful for current business information needs. Some titles were dropped because they were no longer deemed particularly useful; others are no longer being published. This decision process was very much a group project, relying on the expertise of a broad base of business and government publications librarians.

This second edition of *Business Serials of the U.S. Government*, a selected bibliography, is aimed primarily toward the business information needs of librarians in small and medium-sized academic and public libraries, especially those who cannot afford to subscribe to the *American Statistics Index (ASI)*. We feel it will be used mostly by those who are not government documents specialists, although we also believe that documents librarians as well as users in large public and academic libraries and corporate libraries will find it helpful.

One of the major problems we faced in compiling this bibliography was settling on a definition of a serial. The broad definition we finally agreed upon considered a serial to be any title published on a continuing basis. Thus we have included weeklies, monthlies, annuals, biennials, quinquennials, irregulars and virtually every permutation of those frequencies. We have done our best to ensure that all titles are still being published, in some cases even telephoning the responsible issuing agency. We suspect, however, that a few titles may cease before this edition is actually published.

Titles included in this edition are those determined to be primarily business or business-related in nature. For the most part, items of ephemeral value, records of court cases or decisions (except for Internal Revenue publications), agency annual reports (unless truly substantive or primarily statistical), and verifiably inactive titles were excluded. In some cases, inclusion rested upon the judgment of the editors.

Citations include title, initial date of publication (when verifiable), frequency, previous title(s) and dates of publication, special features and Superintendent of Documents (SuDocs) number. When multiple previous titles exist, they have been listed separately on individual lines. In cases of title changes, the initial date of publication given refers to the first publication of which the current title is a direct, linear descendant. Authorities used for bibliographic information were primarily the *Monthly Catalog*, OCLC, and *ASI*. A sample citation is shown in figure 1 with the various items labeled.

Preface vii

Title → **CPI Detailed Report.** Department of Labor. Bureau of Labor Statistics. ← Issuing dept./agency

Initial date of pub. → 1952–   Monthly.   *Consumer Price Index* (1952–74) ← Previous title
                      ↑
                   Frequency

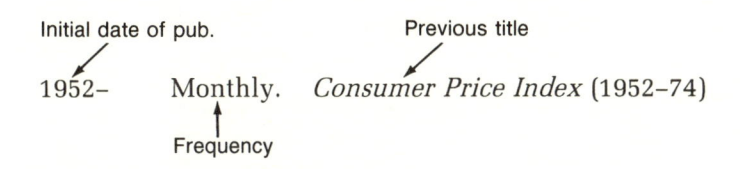

Special features → Tables, charts.   L 2.38/3: ← SuDocs number

Annotation →
Presents monthly price movements for consumer goods. The *Consumer Price Index* measures the average change in prices over time of a fixed-market basket of goods. Indexes are published for two population groups: all urban consumers, and urban wage earners and clerical workers....

Indexed in: *ASI, PAIS*

Figure 1. Sample Citation

The annotations are intended to be evaluative and/or comparative as well as descriptive. Relationships to other titles have been brought out in the annotations. Originally we had hoped to establish a consistent annotation length. However, because of the varied nature of the sources included, especially those serials with extensive sub-series, this proved impossible. At the end of each citation are listed known indexing sources. A list of those indexes, with the abbreviations used, precedes the body of the book. We have also included a separate list of commonly used abbreviations for governmental agencies and the like.

Also provided in this edition are separate title and subject indexes prepared primarily by Bob Rose. Entries are included for previous as well as current titles, and notations are made to those entries where titles are compared as well as to main entries.

viii   Preface

Detailed subject indexing was attempted, using the *Business Periodicals Index (BPI)* for subject authority control. *BPI* was chosen because it is the index most commonly held by our target libraries.

All annotations have been identified with their contributor's initials. The following is a list of those contributors, along with the initials used and institutional affiliation.

PB	Patricia Bick	University of Notre Dame
RB	Rebecca Burke	Arizona State University
DC	Donis Casey	Arizona State University
LH	Lynn Hattendorf	University of Illinois, Chicago
WK	William Kinyon	Texas A&M University
MM	Marcia Meister	University of Colorado, Boulder
SN	Susan Neuman	University of Pittsburgh
JN	Judy Nixon	Purdue University
VS	Virginia Steel	University of California, San Diego
WT	William Taylor	Vanderbilt University
EW	Elizabeth Wood	Bowling Green State University
SW	Steve Wolff	University of Delaware

We would like to make a special note of thanks to the Reference Department staff at the Krannert Graduate School of Management Library for the tremendous amount of work they did in preparing the manuscript and for suffering through our numerous revisions.

PRISCILLA C. GEAHIGAN
Krannert Graduate School of
   Management
Purdue University

ROBERT F. ROSE
The Behrend College
Pennsylvania State University
   at Erie

Co-editors

SELECTED AGENCIES AND OTHER ABBREVIATIONS

BLS	Bureau of Labor Statistics
CAB	Civil Aeronautics Board
CFR	Code of Federal Regulations
CMSA	Consolidated Metropolitan Statistical Area
CPI	Consumer Price Index
CPSC	Consumer Product Safety Commission
DOC	Department of Commerce
DOT	Department of Transportation
EEOC	Equal Employment Opportunity Commission
EIA	Energy Information Administration
EPA	Environmental Protection Agency
FAA	Federal Aviation Administration
FCC	Federal Communications Commission
FDA	Food and Drug Administration
FDIC	Federal Deposit Insurance Corporation
FHLB	Federal Home Loan Bank Board
FSLIC	Federal Savings and Loan Insurance Corporation
GPO	Government Printing Office
ICC	Interstate Commerce Commission
IRS	Internal Revenue Service
MSA	Metropolitan Statistical Area

NTIS	National Technical Information Service
OECD	Organization for Economic Cooperation and Development
OPM	Office of Personnel Management
PMSA	Primary Metropolitan Statistical Area
SBA	Small Business Administration
SEC	Securities and Exchange Commission
SIC	Standard Industrial Classification [Code]
SITC	Standard International Trade Classification
SMSA	Standard Metropolitan Statistical Area
USDA	U.S. Department of Agriculture

ABBREVIATIONS OF INDEX TITLES

ASI	American Statistics Index
BI	Business Index
BPI	Business Periodicals Index
CASSIS	Classification and Search Support Information Systems (online patent database)
F&S	Predicasts F & S Index United States
INFORM	ABI/INFORM (online database)
IUSGP	Index to U.S. Government Periodicals
MC	Management Contents (online database)
PAIS	Public Affairs Information Service Bulletin
RG	Readers' Guide to Periodical Literature
T&I	Trade & Industry Index (online database)
Self	Self indexed

I. GENERAL SOURCES

1. **Budget in Brief.** Executive Office of the President. Office of Management and Budget. 1951– . Annual.
   Tables, charts. PrEx 2.8/2:
Designed for the general public, this document provides a concise, non-technical overview of the budget. Includes the President's budget message and major proposals along with explanation of budget history. Sources of funds are summarized, and uses of funds are discussed with reference to eighteen broad functions performed collectively by various federal programs. Supplements explain the budget and credit-allocation processes and define terms. Tables contrast current and historical data on sources of receipts, expenditures by function and agency, and the relationship of the budget to national debt and credit obligations, gross national product, and the size of the civil service force. (EW)
Indexed in: *ASI*

2. **Budget of the United States Government.** Executive Office of the President. Office of Management and Budget. 1923– . Annual.
   Tables, charts. PrEx 2.8:
Prefaced by the President's budget message and an overview of budget proposals, this document analyzes sources of receipts and eighteen broad categories consuming funds in terms of the national needs served and programs addressing them. A detailed tabulation of the budget by agency and account gives estimates for the current and upcoming fiscal years and actual figures for the previous one. Summary tables of receipts and outlays compare two years of estimates with up to nine years of actual data. Relationship of receipts

2   General Sources

and outlays to the national debt, gross national product, size of the civil service force, and costs of new or incremental programs is shown. Separately published supplements include *Special Analysis of the Budget, Historical Tables, Management of the United States Government*, and a 1,150-page detailed volume entitled *Appendix*, which gives for each agency the texts of appropriation documents, budget breakdowns for each account, relevant legislative proposals, general provisions affecting the agency, and the number of permanent positions allotted to it. Proposed budget supplements and revisions are also shown. (EW)
Indexed in: *ASI*

3. **Bureau of Economic Analysis Staff Papers.** Department of Commerce. Bureau of Economic Analysis. 1963– . Irregular. *Bureau of Economic Analysis Staff Papers BEA-SP* (1963–75)
C 59.14:

These reports present research methodologies and findings in specialized, technical, and less well-established areas of economic analysis. Examples are econometric models, methods of estimating family personal income, and the reliability of the input/output method of calculated national income and product account statistics. (EW)

4. **Census Catalog and Guide.** Department of Commerce. Bureau of the Census. 1790– . Quarterly.
Bureau of the Census Catalog (1974–84)
Bureau of the Census Catalog of Publications (1946–72)
Catalog of U.S. Census Publications (1790–1945)
C 3.163/3:

Designed to facilitate both acquisition and use of census publications, the catalog describes them, gives ordering information, and lists data specialists and abstracts for products released since 1980. Gives title, series, dates of coverage, geographic scope, availability, and principal publication format (print, computer tape, or microform). Earlier products described briefly. *Monthly Product Announcements*, without abstracts, lists new, discontinued, and infrequently published items. Annual cumulation. (EW)

5. **Code of Federal Regulations.** National Archives and Records Administration. Office of Federal Register. 1938– Irregular.
G 4.108:

Established to organize general and permanent regulations of executive departments and agencies of the federal government, this source annually publishes a compilation of regulations, all of which have appeared in daily issues of the *Federal Register* (see item no. 7). Regulations are assigned to one of fifty broad subject areas (titles) corresponding to the titles used in compilations of public laws enacted by Congress. Each volume is revised at least once during the calendar year. The latest amendment to any regulation can be found by consulting either the *CFR*'s list of changes in regulations or the *Federal Register*'s list of such changes. Detailed subject index published as a separate volume. (EW)

6. **County and City Data Book.** Department of Commerce. Bureau of the Census. 1949– . Irregular.
   Maps, charts, tables. C 3.134/2:C83/2:

This *Statistical Abstract* (see item no. 10) supplement, overlapping somewhat with the *State and Metropolitan Area Data Book* (see item no. 9), covers states, counties, metropolitan statistical areas, cities of 25,000 people and over, and places of 2,500 or more inhabitants. Most information drawn from latest available census of population and housing, government, manufactures, wholesale and retail trade, and service industries, supplemented with data from other government and private organizations. Rankings given for metropolitan statistical areas (MSAs) on selected data items. Appendix presents data on components of MSAs and on New England county metropolitan areas. (EW)
Indexed in: *ASI*

7. **Federal Register.** National Archives and Records Administration. Office of the Federal Register. 1936– . Daily.
   GS 4.107:

Established to disseminate regulations and legal notices of federal departments and agencies, this source publishes such regulations along with agency policy statements, statements of organization and function, presidential proclamations, and executive orders. The *Register*'s legal functions are (1) providing official notice that such documents exist; (2) providing opportunity for public comments; (3) making available the full text; and (4) indicating the date of issuance of final regulations. Various finding aids (e.g., list of dates when regulations become effective) in each issue. Codified (arranged numerically) annually in the *Code of Federal Regulations* (see item no. 5). (EW)
Indexed in: *CFR*, Self

4 General Sources

8. **Standard Industrial Classification Manual.** Executive Office of the President. Office of Management and Budget. 1939–   . Irregular.
PrEx 2.6/2:In27/

Developed to promote comparability of business and industrial statistics, this manual features a system for classifying business establishments by the primary type of activity in which they engage. Organized into eleven major industry divisions, each assigned a 2-digit code, it identifies further levels of progressively more specific industry subdivisions within the basic eleven divisions. Each subdivision is assigned the appropriate 2-, 3-, or 4-digit code to distinguish it from other sectors of the industry. The standard for federal data collection and analysis, this manual is widely used by business as well. Latest edition is 1987. (EW)

9. **State and Metropolitan Area Data Book.** Department of Commerce. Bureau of the Census. 1971–   . Irregular.
Maps, tables. C 3.134/5:

A *Statistical Abstract* (see item no. 10) supplement, this source presents data from nearly fifty federal and private agencies. Most information based on the latest available Censuses of Population and Housing, Government, Manufactures, Retail and Wholesale Trade, and Service Industries. State, county, and metropolitan statistical areas (MSAs) covered. MSAs ranked on various demographic and economic measures. Larger ones separated into primary (PMSAs) and consolidated metropolitan statistical areas (CMSAs). Data for central cities of MSAs given. More trend data than *County and City Data Book* (see item no. 6), but there is some overlap. Appendixes for terms and concepts, alphabetical listings of central cities and metropolitan counties, and additional New England data. (EW)
Indexed in: *ASI*

10. **Statistical Abstract of the United States.** Department of Commerce. Bureau of the Census. 1878–   . Annual.
Tables, maps, charts. C 3.134:

The standard compendium for U.S. social, political, and economic statistics for well over a hundred years, this source serves as both a reference for statistics and a guide to more detailed government and private statistical sources. Emphasis is on national statistics. Considerable data for regions and states; less data for metropolitan areas and cities. Appendixes give cross references to *Statistical Abstract* supplements, discuss methodology and sources of statistics, and give state rankings for selected demographic and economic measures. Detailed subject index. Separately published supplements include *County and City Data Book* (see item no. 6), *State and Metropolitan Area Data Book* (see item no. 9), *Congressional District Data Book*, and *Historical Statistics of the United States*. (EW)

11. **United States Government Manual.** National Archives and Records Administration. Federal Register Office. 1935– . Annual. *United States Government Organizational Manual* (1935–73).
    Text of the Constitution and Declaration of Independence. GS 4.109:

A special edition of the *Federal Register*, this manual covers federal agencies and programs. Directory information is given together with a description of the mission and often a brief history, explanation of the scope of activity, the size, and a statement of philosophy. Focus is on legislative, executive, and judicial branches of government. Limited information given for independent, multinational, quasi-official, and other types of government organizations. Appendixes list defunct agencies and functions, abbreviations and acronyms, organization charts, standard federal regions, and agencies appearing in the *Code of Federal Regulations*. Name and subject/agency indexes. (EW)

II. ECONOMIC CONDITIONS

12. **Annual Statistical Digest [Federal Reserve System. Board of Governors].** Federal Reserve System Board of Governors. 1971/75– . Annual.
Tables. FR 1.59:

Contains economic and financial time series for statistical data contained in the *Federal Reserve Bulletin* (see item no. 18). Also includes domestic nonfinancial series for which the Board of Governors is the primary source. Maintains historical series published in *Banking and Monetary Statistics, 1949–1970*. Provides information about money supply, interest rates, bank assets and liabilities, mortgages, flow of funds, installment credit, personal income and savings, economic indicators, government securities and bonds, corporate profits, and business credit. (SN)

13. **Business Conditions Digest.** Department of Commerce. Bureau of Economic Analysis. 1961– . Monthly.
B.C.D. (Nov. 1968–Dec. 1971)
Business Cycles Development (1961–68)
Tables, charts, historical series. C 59.9:

Presents monthly economic time series useful to business analysts and forecasters. Part I reports cyclical indicators including composite indexes; employment and unemployment; production and income; consumption, trade orders, and deliveries; fixed capital investment; prices, costs, and profits; money, credit, and diffusion indexes. Part II contains tables on national income and product; prices, wages, and productivity; labor force, employment, and unemployment; government activities; U.S international transactions; and

Economic Conditions 7

international comparisons. Part III includes appendixes of historical data and special measures and factors. (SN)
Indexed in: *ASI, IUSGP, PAIS*

14. **Business Statistics.** Department of Commerce. Bureau of Economic Analysis. 1951– . Biennial. *Statistical Supplement to the Survey of Current Business* (–1949)
    Tables, historical series. C 59.11/3:

Presents monthly and annual data that are cumulated from the monthly *Survey of Current Business* (see item no. 25). Also provides methodological notes. Includes data on general business indicators, commodity prices, construction and real estate, domestic trade, labor force, employment and earnings, finance, foreign trade, transportation and communication, chemicals and allied products, electric power and gas, food and kindred products, metals and manufactures, petroleum, coal, leather, pulp, paper products, rubber and rubber products, stone, clay and glass products, textile products, and transportation equipment. (SN)
Indexed in: *ASI*

15. **CPI Detailed Report.** Department of Labor. Bureau of Labor Statistics. 1952– . Monthly. *Consumer Price Index* (1952–July 1974)
    Tables, charts. L 2.38/3:

Presents monthly price movements for consumer goods. The Consumer Price Index measures the average change in prices over time of a fixed market basket of goods. Indexes are published for two population groups: all urban consumers, and urban wage earners and clerical workers. Separate indexes are also published for selected local areas. Tables for U.S. cities include expenditure categories, seasonally adjusted expenditure categories, food expenditure categories, nonfood expenditure categories, and seasonally adjusted nonfood expenditure categories. (SN)
Indexed in: *ASI, PAIS*

16. **Economic Indicators.** Prepared for the Joint Economic Committee by the Council of Economic Advisors. May 1948– . Monthly.
    Tables, charts. Y 4.Ec7: Ec7/

Presents charts and tables on economic statistics such as gross national product, GNP implicit price deflator, national income, personal consumption expenditure, corporate profits, expenditures for

new plants and equipment, employment, unemployment, wages, industrial production and capacity utilization, new construction, business sales and inventories, manufacturers' shipments, inventories and orders, prices, money, credit, and security markets—including interest rates, federal finance, and international statistics. Collects major economic statistical indicators in one monthly publication for the edification of Congress and the public. Has two supplements: *Supplement to Economic Indicators* and *Descriptive Supplement to Economic Indicators.* (SN)
Indexed in: *ASI, PAIS,* T&I

17. **Economic Report of the President Transmitted to the Congress.** Executive Office of the President. 1950– . Annual. *Economic Report of the President to the Congress* (1947–49) Tables. Y 1.1/7:

Contains the economic report as transmitted by the President to Congress. Discusses current economic situations and problems. Contains the *Annual Report* of the Council of Economic Advisors. Statistical Appendix includes data on income, employment, and production. Provides information on national income and expenditure; population, employment, wages, and productivity; production and business activity; prices; money supply, credit and finance; government finance; corporate profits and finance; agriculture and international statistics. The reports for 1950–53 include *The Annual Economic Review* by the Council of Economic Advisors; for 1954 it includes the Council's *Annual Economic Report.* (SN)
Indexed in: *ASI*

18. **Federal Reserve Bulletin.** Federal Reserve System. Board of Governors. May 1915– . Monthly.
Tables, charts. FR 1.3:

Contains announcements and articles regarding the actions of the Board of Governors. Also contains the minutes of meetings of the Federal Open Market Committee. Includes section on financial and business statistics. Provides tables on money supply and bank credit, policy instruments (including interest rates), federal reserve banks, monetary and credit aggregates, commercial banking institutions, financial markets, federal finance, securities markets and corporate finance, real estate, consumer installment credit, flow of funds, international statistics, and international interest and exchange rates. (SN)
Indexed in: *ASI, BPI, F&S,* INFORM, *IUSGP,* MC, *PAIS*

Economic Conditions 9

19. **Federal Reserve Chart Book.** Federal Reserve System. Board of Governors. 1976– . Quarterly. *Monthly Chart Book* (1976–78)
Charts. FR 1.30:

Presents charts describing statistics on banking and money supply. Includes charts on reserves, money stock, growth of monetary and banking aggregates, income velocity of money, deposit turnover, economic activity, funds raised and supplied, federal finance, state and local finance, corporate security issues, nonfinancial corporations, bank loans to businesses, household finance, mortgage debt and construction, financial institutions, the stock market and interest rates, U.S. international transactions, and foreign interest and exchange rates. Many of the statistics charted are produced in tabular form in other Federal Reserve publications. Supplemented by *Historical Chart Book* (see item no. 21). (SN)
Indexed in: *ASI, PAIS*

20. **Handbook of Cyclical Indicators.** Department of Commerce. Bureau of Economic Analysis. May 1977– . Irregular.
Tables. C 59.9/3:In2/

Collects information for about 300 time series originally published in *Business Conditions Digest* (see item no. 13). Emphasis on cyclical indicators as defined by the Bureau of Economic Analysis in cooperation with the National Bureau of Economic Research in 1975. Publication in four parts: Part I, Series Descriptions; Part II, Composite Indexes of Leading, Coincident, and Lagging Indicators; Part III, Historical Data, 1947–82; and Part IV, Reference Materials. Also includes a Bibliography, a list of Source Agencies, and Series Finding Guides. (SN)
Indexed in: *ASI*

21. **Historical Chart Book.** Federal Reserve System. Board of Governors. 1965– . Annual.
Charts. FR 1.30/2:

Presents long-range financial time series in a chart format. Contains series not included in *Federal Reserve Chart Book* (see item no. 19) as well as many that appear there. Charts include monetary and reserve aggregates, measures of economic growth, gross national product and business investment, labor force, income and earnings, industrial production, prices, debt and borrowing, federal, state and local sector data, corporate sector data, household finance, mortgage debt and construction, commercial banks, reserve banks, financial

10   Economic Conditions

institutions, stock market, interest rates, and international transactions. Issued as a supplement to *Monthly Chart Book* (1977-78). (SN)
Indexed in: *ASI*

22. **Local Area Personal Income.** Department of Commerce. Bureau of Economic Analysis. 1969/74-   . Annual.
C 59.18:

Presents estimates of total and per capita personal income for counties and metropolitan areas. Summaries for national, regional, and state estimates are included in Volume 1. This volume also contains the methodology statement and tables on the distribution and changes in total personal income for regions. Volumes 2-9 contain estimates for one of the eight Bureau of Economic Analysis regions: New England, Midwest, Great Lakes, Plains, Southeast, Southwest, Rocky Mountain, and Far West. (SN)
Indexed in: *ASI*

23. **National Income and Product Accounts of the United States.** Department of Commerce. Bureau of Economic Analysis. 1965-   . Irregular. *National Income and Product Accounts of the United States Statistical Tables* (1965, 1974)
Tables. C 59.11/4:In2/

Presents estimates that resulted from the comprehensive benchmark revision of January 1976. Provides information on national outputs and receipts over time (1929 on). Includes tables on gross national product, net national product, and national income; personal income and outlay; government receipts and expenditures; foreign transactions; saving and investment; product, income, and employment by industry; and implicit price deflators and price indexes. Supplement to *Survey of Current Business* (see item no. 25). (SN)
Indexed in: *ASI*

24. **Social Security Bulletin.** Department of Health and Human Services. Social Security Administration. 1938-   . Monthly.
HE 3.3:

Presents statistics and analyses of social security and related programs. Includes articles about these programs and relevant legislation in the United States and abroad. Provides monthly tables on the Income-Maintenance Program; Social Security Trust Funds; Old-Age, Survivors, and Disability Insurance Cash Benefits; Supplemental Security Income; Public Assistance; Black Lung Benefits; Unem-

ployment Insurance; and Economic Indicators. Publishes an annual statistical supplement. (SN)
Indexed in: *ASI, BPI*, INFORM, *IUSGP, PAIS*

25. **Survey of Current Business.** Department of Commerce. Bureau of Economic Analysis. 1921-   . Monthly. *Monthly Supplement of the Commerce Reports* (July 1921–July 1925) C 59.11:

Monthly statistical guide to the U.S. economy. Includes reviews of business conditions as well as articles on various economic statistical time series, such as personal income by state, international transactions, manufacturing and trade inventories and sales. Section on blue paper gives current business statistics. Topics covered include fixed capital stock, gross national product, government transactions, input-output, inventories and sales, national income and product accounts, plant and equipment expenditures, pollution abatement and control, profits, foreign investments, balance of payments, commodity prices, employment and unemployment, transportation, and various statistics by specific industry such as food or chemicals. (SN)
Indexed in: *ASI, BPI, F&S, IUSGP*, INFORM, MC, *PAIS*

III. DEMOGRAPHICS

26. **Census of Population.** Department of Commerce. Bureau of the Census. 1790– . Decennial.
C 3.223/

Contains detailed statistical data on the characteristics of the population such as age, sex, race, education, occupation, and income by state, county, city, and town. Volume 1, *Characteristics of the Population*, contains the following chapters: Number of Inhabitants (C3.223/5:), General Population Characteristics (C3.223/6:), General Social and Economic Characteristics (C3.223/7:), and Detailed Population Characteristics (C3.223/8:). Volume 2 is *Subject Reports*, each of which focuses on a particular topic with very detailed data. Some of the titles are: Earnings by Occupation and Education, Occupation by Earnings, and Place of Work. (PB)
Indexed in: *ASI*

27. **Census of Population and Housing.** Department of Commerce. Bureau of the Census. 1790– . Decennial.
SUDOCS no.: see annotation

Issued in four series, the most useful of which are Block Statistics (C3.224/5:), which provides data on population, age, race, home ownership, and median value of home by individual city blocks, and Census Tracts (C3.223/11:), which provides such socioeconomic data as age, sex, race, income, and occupation by individual census tracts. Also issued are: Summary Characteristics for Governmental Units and Standard Metropolitan Statistical Areas (C3.223/23:) and Congressional Districts of the —th Congress (C3.223/20:). (PB)
Indexed in: *ASI*

12

28. **Current Population Reports.** Department of Commerce. Bureau of the Census. Varies. Irregular.
SUDOCS no.: see annotation

Included in this series are the following subseries: Population Characteristics (C3.186/17:P-20) provides data on year of schooling by age, sex, and race, fertility of American women, and mobility of the population; Special Studies (C3.186:P-23); Population Estimates and Projections (C3.186:P-25) provides data such as population projections through the year 2000, monthly growth rates, and estimates of the population by age, sex, and race; Federal-State Cooperative Programs for Population Estimates (C3.186:P-26); Farm Population (C3.186:P-27); Special Censuses (C3.186:P-28); Consumer Income (C3.186/16:P-60) contains data on money income of households, the poverty level, and persons receiving non-cash benefits; Household Economic Studies, Economic Characteristics of Households in the U.S. (C3.186:P-70). (PB)
Indexed in: *ASI*

29. **Data User News.** Department of Commerce. Bureau of the Census. 1966– . Monthly.
Small-Area Data Indicators (1966–68)
Small-Area Data Notes (1969–74)
C 3.238:

A "current awareness" publication covering recent Census Bureau reports; forthcoming publications in print, micro, or computerized format; and changes in means of collecting data or types of data collected. (PB)
Indexed in: *IUSGP*

30. **Monthly Vital Statistics Report.** Department of Health and Human Services. National Center for Health Statistics. 1952– . Monthly.
Current Mortality Analysis (1943/44–51)
Monthly Marriage Report (–1952)
Monthly Vital Statistics Bulletin (1938–52)
HE 20.6217:

Provides data on births, marriages, divorces, and deaths for the current month, but only on a state-by-state basis. Also issues supplements containing advance reports of final annual cumulative statistics. Data are cumulated in *Vital Statistics of the United States* (see item no. 32). Annual summary appears in issue number 13. (PB)
Indexed in: *ASI*

14   Demographics

31. **U.S. Decennial Life Tables.** Department of Health and Human Services. National Center for Health Statistics. 1890–   . Annual. *U.S. Life Tables* [and other variations] (1890–1975)
Tables. HE 20.6215:

Issued in two volumes annually. Volume I is in four parts: U.S. life tables; U.S. life tables, eliminating certain causes of death; methodology of the national and state life tables; and some trends and comparisons of U.S. life table data, 1900–81. Volume II contains life tables for each state, issued in 50 parts. (PB)
Indexed in: *ASI*

32. **Vital Statistics of the U.S.** Department of Health and Human Services. National Center for Health Statistics. 1937–   . Annual.
Tables. HE 20.6210:

Issued in three volumes: natality, mortality, and marriage and divorce. This is the definitive source for extensive basic data and analysis on marriage, divorce, natality, fetal mortality, and mortality. It is not, however, as current as most would prefer. The 1981 data appeared in 1985. More current data is available in *Monthly Vital Statistics Report* (see item no. 30) and *Advance Data from Vital and Health Statistics* (HE 20.6209/3). (PB)
Indexed in: *ASI*

IV. INTERNATIONAL BUSINESS

33. **Area Handbook Series.** Department of the Army. 1962–   . Irregular.
    Maps, tables, charts, graphs, bibliography, illustrations, glossary. D 101.22:550:

Handy reference series covering more than 140 countries with one volume per country. The analyses are written by teams of social scientists and describe the economy and political and social systems of each country. Although the volumes are not updated regularly, they provide a broad background and overview of each country and are helpful for anyone beginning research on a geographic area. Detailed subject index at end of each volume. Formerly titled *Area Handbook for [Country]*, individual volumes are now titled *[Country]: A Country Study*. (VS)

34. **Background Notes.** Department of State. 1954–   . Irregular. Map, bibliography. S 1.123:

Series covering more than 150 countries providing very brief information on each one. Short sections on the geography, people, history, government, political conditions, economy, and foreign relations are included with a brief bibliography for further information. Travel notes are also provided. This is a useful series for ready reference purposes. Approximately 75 of the country reports are updated each year. (VS)

35. **Business America.** Department of Commerce. 1970–   . Biweekly.
    Commerce America (1976–78)
    Commerce Today (1970–75)

15

16   International Business

International Commerce (1940-70)
C 61.18:
Periodical featuring brief articles on foreign trade topics. Regular columns on the business outlook abroad and foreign market briefs provide current information on the trade climate of approximately 12 countries. Also included are book reviews, a calendar for world traders, and lists of overseas business opportunities. Every six months feature articles discuss recent trends and events in world trade, and the outlook for the future. (VS)
Indexed in: *ASI, BI, BPI,* INFORM, *IUSGP,* MC, T&I, Self

36. **Country Market Survey (CMS).** Department of Commerce. International Trade Administration. Irregular.
Maps, charts, graphs, tables. C 61.9:
Useful series of brief reports focusing on a specific country and class of products. Information related to the market for the product is provided and includes an assessment of the competition, end users, and market practices. Also included is a brief section discussing the economy of the country, including some statistics, as well as a list of agencies and organizations to contact for additional information. Citations to other related International Trade Administration publications are provided. Covers 18 product classes and 40 countries. (VS)
Indexed in: *ASI*

37. **Customs Regulations of the United States.** Department of the Treasury. Customs Service. Continuing supplements cumulated irregularly.
T 17.9:
Reprints of material from the *Code of Federal Regulations* (see item no. 5) covering the rules and regulations relating to exports and imports. Includes information about products, records to be maintained, forms required by the U.S. Customs Service, and rates of duty. The subject index at the end of the volume facilitates identification of the relevant sections. (VS)

38. **Export Briefs.** Department of Agriculture. Foreign Agricultural Service. 1979-   . Weekly.
A 67.40/2:
News publication listing agricultural products wanted by foreign buyers. Includes products, quantity, quality, packaging, delivery, and who to contact. Also includes market briefs for specific commodities in specific countries, and lists of upcoming food shows or

exhibitions. Good current awareness service of export opportunities for agribusiness. (VS)

39. **Foreign Economic Trends and Their Implications for the United States.** Department of Commerce. International Trade Administration. 1969– . Irregular (annual or semiannual for each country). *Economic Trends and Their Implications for the United States*
Tables. C 61.11

Country-specific series of reports prepared by American embassy staff personnel describing the economies of individual countries. Although there is some variation, the data provided usually include key economic indicators and discussions of the current economic situation and outlook, money and banking, trade, balance of payments, labor and wages, and the potential effect those items may have on the United States. For anyone seeking a brief overview of a country's economy, these reports provide easily readable analyses and summary statistics. The reports average 10 to 25 pages. (VS)

40. **Highlights of U.S. Export and Import Trade.** Department of Commerce. Bureau of the Census. 1967– . Monthly. *U.S. Foreign Trade: Highlights of Exports and Imports*
Tables. C 3.164:990/

Detailed statistical reports of U.S. exports and imports having values of $1,000 or more. Data are provided by month, year, type of product (including both general groupings and more specific industry breakdowns), and geographic locations (including country and regional data). This is a fairly detailed source of import and export statistics that is useful for reference purposes. This series is more widely known and frequently requested than *U.S. Foreign Trade Highlights* (see item no. 49). (VS)
Indexed in: *ASI*

41. **International Finance: Annual Report to the President and to the Congress.** National Advisory Council on International Monetary and Financial Policies. 1973/74– . Annual.
Tables. Y 3.N21/16:1/

Text of report summarizing U.S. situation with respect to the balance of payments and the international monetary climate. Describes U.S. involvement with the International Monetary Fund, World Bank, and other economic assistance programs. Various appendixes provide statistical data related to international finance as well as descriptions of loans and other forms of assistance given to countries by multilateral development banks. For those following

18   International Business

the international debts situation, this report is an important annual summary of U.S. involvement. (VS)

42. **Key Officers of Foreign Service Posts.** Department of State. 1964–   . Three times per year (Quarterly until 1972). *Key Officers of Foreign Service Posts Guide for Businessmen* S 1.40/5:

Small booklet that briefly explains types of positions found in foreign missions and their potential usefulness to U.S. business people; also lists U.S. foreign service posts. Addresses, phone numbers, and names of key personnel are provided. Concise instructions on addressing mail to foreign service posts and a geographical index are also included. (VS)

43. **Overseas Business Reports.** Department of Commerce. International Trade Administration. 1962–   . Irregular.
    Tables, bibliography. C 61.12:

Each report in this series is devoted to a specific country and offers marketing advice for U.S. business people. Current information and statistics summarize the state of the economy and the foreign trade outlook. An extensive section on trade regulations provides instructions on investing, licensing, and marketing products in the country. Names and addresses of government agencies responsible for foreign trade and for U.S. agencies and organizations are listed, as is a brief guide for business travelers. This is a handy series for anyone wishing to do business in a foreign country. (VS)
Indexed in: *ASI*

44. **Post Reports.** Department of State. Irregular.
    Illustrations, maps, bibliography. S 1.127:CT

Each report in this series covers one country and provides brief information on the local geography, climate, population, and political and social institutions. The reports also include detailed descriptions of the local U.S. embassy, including photographs. A section of notes for travelers increases the value of this series as a useful guide for people traveling to a country for business or pleasure. (VS)

45. **State Export Series.** Department of Commerce. International Trade Administration. 1977–   Very irregular.
    Tables, graphs, charts. C 61.30:CT

Set containing 50 parts, one for each state, with titles such as *Minnesota Exports*. The statistics provided are estimates of agricultural, manufacturing, and other types of exports compiled from reports

filed by each state with the federal government. Also includes comparisons to other states and rankings of all 50 states in each part. This service has been published twice in the past 9 years. Earlier ones under C 57.29:. (VS)
Indexed in: *ASI*

46. **Tariff Schedules of the United States, Annotated.** International Trade Commission. 1963–   . Annual (irregular before 1980).
    ITC 1.10:

Lengthy publication containing the legal text of the U.S. Tariff Schedules. Most of the volume is devoted to an enumeration of articles detailing all U.S. exports and imports so that importers may determine the classifications and duty rates for articles to be imported. Requirements for statistical reports to the government are also included. The tariff schedules are classified as follows: animal and vegetable products, wood, paper and printed matter, textile fibers and products, chemicals and related products, non-metallic minerals, and metals and metal products. Can be difficult to use if one is not familiar with the export regulations. (VS)
Indexed in: *ASI*

47. **TOP Bulletin: A Joint Activity of the U.S. Department of Commerce and the U.S. Foreign Service. U.S. Department of State.** Department of Commerce. Trade Opportunities Program. 1978–   . Weekly.
    C 61.13:

Current announcements of export opportunities available to U.S. businesses, gathered daily by personnel at U.S. foreign service posts. Several types of opportunities are included: (1) direct sales leads from prospective overseas private sector buyers, (2) agent/distributorship opportunities, (3) foreign government bid invitations, and (4) notices of foreign buyer visits to the United States, as well as overseas trade missions, catalog shows, trade fairs, and exhibitions. A valuable up-to-date source for anyone interested in trading in foreign countries. (VS)

48. **U.S. Department of State Indexes of Living Costs Abroad, Quarters, Allowances, and Hardship Differentials.** Department of Labor. Bureau of Labor Statistics. 1979–   . Quarterly.
    U.S. Department of State Indexes of Living Costs Abroad and Quarters Allowances (1979–82)
    L 2.101:

20   International Business

Cost-of-living indexes computed in order to compensate American foreign service employees for higher costs of living in different locations. Data include exchange rates and cost-of-living indexes by country, and sometimes by major city within a country. The explanatory notes contain the following caveat: "The indexes should not be used to compare living costs of foreign nationals living in their own country, since the indexes reflect only the expenditure pattern and living costs of Americans." (VS)
Indexed in: *ASI*

49. **U.S. Foreign Trade Highlights.** Department of Commerce. International Trade Administration. 1984–   . Semiannual.
    Tables, charts, graphs. C 61.28/2:

In-depth statistics on U.S. import and export activity with major trading partners (by country) and categories (developed countries, developing countries, and centrally planned economies). Shows trends for several years, growth rates, and composition of exports and imports. Includes rankings of top 50 U.S. export markets, top 50 foreign suppliers, and top 50 U.S. foreign trade partners, as well as a world trade overview. This may be more detailed than necessary for some libraries. (VS)
Indexed in: *ASI*

50. **U.S. Import and Export Price Indexes.** Department of Labor. Bureau of Labor Statistics. Quarterly.
    Tables. L 2.60/3:

Press release containing text summary of export and import price indexes for the following categories: food, beverages, and tobacco; crude materials; fuels and related products; chemicals; manufacturing products and machinery; and transportation equipment. Tables provide detailed statistics arranged by SITC number and including index base, index figures for past two quarters, and percent changes in annual and quarterly indexes. A specialized reference source that is valuable for current international trade statistics. (VS)
Indexed in: *ASI*

51. **United States Trade Performance in [Years] and Outlook.** Department of Commerce. International Trade Administration. 1983–   . Annual.
    Tables, charts, graphs. C 61.28:

Valuable overview of U.S. foreign trade situation. Includes articles on topics relevant to recent trade developments and numerous statistical tables that provide data in export and import activity. This is an informative source of fairly current information on the U.S. situation in world trade, and the statistics presented include the most frequently requested aggregates without being overly detailed. (VS) Indexed in: *ASI*

V. INDUSTRY

A. General

52. **County Business Patterns.** Department of Commerce. Bureau of the Census. Varies. Annual.
    C 3:204/3:

A series of annual reports providing data on detailed economic activity at the county level for the United States, the 50 states, and the District of Columbia. Data are available by SIC code on employment, payroll, and number and employment-size of establishments for private nonfarm organizations. Excludes data on government employees, self-employed persons, farm workers, and domestic workers. Each report treats one state and one report summarizes the data for the United States. (SN)
Indexed in: *ASI*

53. **Current Industrial Reports.** Department of Commerce. Bureau of the Census. Varies. Most are monthly with annual cumulations.
    C 3.158: (for most reports)

A series of reports on separate commodity surveys issued monthly, quarterly, annually, or biennially depending on the survey. Data are collected for 5,000 products and cover much of industrial activity. Industries include the major 2-digit SIC groupings such as apparel and leather; chemicals, rubber, and plastics; metal products; lumber, furniture, and paper products; machinery and equipment; processed foods; stone, clay, and glass products; and textile mill products. Reports include titles such as: Tractors; Consumption on

the Cotton System and Stocks; Women's and Children's Outerwear; Pollution Abatement Costs and Expenditures. (SN)
Indexed in: *ASI*

54. **Quarterly Financial Report for Manufacturing, Mining and Trade Corporations.** Department of Commerce. Bureau of the Census. 4th Quarter, 1974– . Quarterly. *Quarterly Financial Report for Manufacturing Corporations* (3rd Quarter 1955–3rd Quarter 1974)
    FT 1.19:

Presents aggregate statistics on the financial results and position of American corporations. Uses a sample survey to determine estimated statements of income and retained earnings, balance sheets, and financial and operating ratios for manufacturing, mining, and trade corporations. Data are classified by industry and by asset size. Industries are classified by the Enterprise Standard Industrial Classification (ESIC). Excludes foreign subsidiaries and domestic companies primarily engaged in banking, finance, or insurance. Tables include rates of change in sales and profits, profits per dollar of sales, annual rates of profit on stockholders' equity, and rates of return. Issued 1975–82 by the Federal Trade Commission. (SN)
Indexed in: *ASI*

55. **U.S. Industrial Outlook.** Department of Commerce. Bureau of Industrial Economics. 1960– . Annual. *U.S. Industrial Outlook for... Industries with Projections for...* (1960–83)
    C 61.34:

Provides a summary of activity and outlook for over 350 service and manufacturing industries. For the most part follows the SIC system. Includes major industries such as construction, electric lighting and wiring equipment, wood products, pulp, paper and board, metal cans, glass containers and plastic bottles, mining, coal, petroleum and natural gas, chemicals, rubbers and plastics, coatings and adhesives, drugs, industrial machinery, computer equipment and software, telephone and telegraph equipment and services, medical and dental instruments, and many others. (SN)
Indexed in: *ASI*

24 Industry

B. Banking and Finance

56. **Receivables Outstanding at Finance Companies.** Federal Reserve System. Board of Governors. Monthly.
    Sales Finance Companies
    Consumer Credit Outstanding at Consumer Finance Companies
    Finance Companies
    Tables. FR 1.26:

Gives the amount of receivables at end of each month, and gives changes for outstanding credit held, credit extended, and number of vehicles financed by sales finance companies. A separate table gives data on total credit sales of new passenger cars as a percent of total number sold at retail. Figures on amounts outstanding, extensions, and repayments later appear in the *Federal Reserve Bulletin* (see item no. 18). Changed to the current title in February, 1985. (SW)
Indexed in: *ASI*

57. **SEC Monthly Statistical Review.** Securities and Exchange Commission. 1942– . Monthly. *Statistical Bulletin* (1942–79)
    Tables. SE 1.20:

Regularly appearing data include stock market statistics, option market statistics, and security registration statistics. Other special features appear on a semiannual or annual basis. The textual Statistical Highlights provides a quick overview of the data. (SW)
Indexed in: *ASI*

58. **Savings and Home Financing Sourcebook.** Federal Home Loan Bank Board. 1952– . Annual.
    Tables. FHL 1.11:

Provides statistical tables on savings and home financing, with emphasis placed on the institutions affiliated with the FHLB system and FSLIC. Kept up to date by periodic reports published by the Board. Continues Federal Home Loan Bank Board *Statistical Summary*. (SW)
Indexed in: *ASI*

59. **Treasury Bulletin.** Department of Treasury. 1939– . Quarterly. *Bulletin of the Treasury Department* (1939–June 1945)
    Tables, charts, graphs. T1.3:

An excellent source for a broad range of statistics on the U.S. government, and information relative to the operations of the Treasury. The *Bulletin* is divided into four parts: (1) Financial Operations;

(2) International Statistics; (3) Cash Management/Debt Collection; and (4) Special Reports. Issued monthly, 1939–82. (SW)
Indexed in: *ASI*

C. Communications

60. **Annual Report.** Federal Communications Commission. 1935– . Annual.
    Tables, charts, graphs. CC 1.1:

An all-in-one source for yearly activity in the areas of legislation, litigation, and enforcement; policy planning, analysis, and research; broadcast services, broadcast and cable TV statistics, and financial data; and common carrier and private radio statistics. Biographical sketches of FCC commissioners from 1934 to the present and an FCC Organizational Chart are included. Essential reference source for keeping up with the ongoing activities and innovations in the communications field. (LH)
Indexed in: *ASI*

61. **Major Matters Before the Federal Communications Commission.** Federal Communications Commission. 1964– . Annual.
    CC 1.50:

Presents the Commission's accomplishments for the calendar year and provides a description of the central issues involved in major matters before the Commission. Provides detailed discussions of major matters in progress, by area (8), gives docket numbers, staff contacts, and relevant references. Excellent source for precise information that allows one to trace the history of an issue. (LH)

62. **Statistics of Communication Common Carriers.** Federal Communications Commission. 1939– . Annual. *Statistics of the Communications Industry in the United States* (1939–57)
    Charts, tables. CC 1.35:

Financial and operating data taken from annual and monthly reports submitted to the FCC by all common carriers engaged in interstate or foreign communication service. Information such as the number of telephones in the United States is included, along with developments, revenues, overseas communication service, and employee information. Super source for finding information about specific utility companies in addition to Moody's. Company and subject indexes. For quarterly data see *Quarterly Operating Data of*

26 Industry

Telephone Carriers (CC 1.14) and *Quarterly Operating Data of Telegraph Carriers* (CC 1.13). (LH)
Indexed in: *ASI*

D. Construction

63. **Annual Housing Survey: United States and Regions.** Department of Commerce. Bureau of the Census. 1973– . Biennial. Tables. C 3.215:H-150-

The survey reports cover current data on housing characteristics, detailing items such as home financing, the quality of the neighborhood, and the delivery of public services. Arranged in six parts, the survey consists of: Part A, General Housing Characteristics; Part B, Indicators of Housing and Neighborhood Quality; Part C, Financial Characteristics of the Housing Inventory; Part D, Housing Characteristics of Recent Movers; Part E, Urban and Rural Housing Characteristics U.S. and Regions; and Part F, Energy-Related Housing Characteristics. Each part summarizes the nation, and the Northeast, Midwest, South, and West. (WT)
Indexed in: *ASI*

64. **Census of Construction Industries.** Department of Commerce. Bureau of the Census. 1967– . Quinquennial.
    C 3.245:

Part of the economic census. Published in years ending in "2" or "7" (e.g., 1982). Provides a detailed analysis of all construction establishments with a payroll. These establishments primarily engage in contract construction, developing, or land subdividing. The detailed analysis includes the number of establishments, payroll, assets, and type of construction—further divided into industry, states and SMSAs. The 1982 Construction Census consisted of 28 reports in the Industry Series (/3:CC 82-I-), 10 reports in the Geographic Series (/7:CC 82-A-), and a special report, Legal Form of Organization and Type of Operation (/5:CC 82-SP-1). (WT)
Indexed in: *ASI*

65. **Construction Reports.** Department of Commerce. Bureau of the Census. Varies. Monthly/Quarterly.
    Tables, charts, graphs. C 3.215/

Reporting a full statistical range on the construction of housing units, the eight subseries cumulate data collected by the Construction Statistics Division. The subseries include Housing Starts

(/2:C20-, monthly, 1959–), New Residential Construction in Selected Metropolitan Statistical Areas (/15:C21-, quarterly, 1973–), Housing Completions (/13:C22, monthly, 1970–), New One-Family Houses Sold and For Sale (/9:C25, monthly, 1962–), Price Index of New One-Family Houses Sold (/9-2:C27, quarterly, 1974–), Value of New Construction Put in Place (/3:C30-, monthly, 1959–), Housing Units Authorized by Building Permits (/4:C40-, monthly, 1980–), Residential Alterations and Repairs (/8:C50-, quarterly, 1961–). (WT)

66. **Construction Review.** Department of Commerce. International Trade Administration. 1955– . Bimonthly.
    Tables. C 62.10:

Each issue contains a feature article and a statistical series. Articles cover a current topic focusing on an aspect of the industry. Compiling the statistical series from governmental and private sources, the series details data for the construction of public and private, residential and nonresidential building. The data include such items as building permits, contract awards, construction materials, and contract construction employment. (WT)
Indexed in: *BI, BPI, F&S, IUSGP, PAIS,* T&I

E. Manufacturing

67. **Annual Survey of Manufactures.** Department of Commerce. Bureau of the Census. 1949– . Annual.
    Tables. C 3.24/9-9:

This annual survey is conducted for each of the years between the Censuses (see item no. 69), and serves as an interim source of data. Key measures of manufacturing activity for industries and industry groups, as well as other detailed statistics on manufacturing activity, are provided. Data are presented in five major categories: (1) statistics for industry groups and industries; (2) value of product shipments; (3) value of manufacturers' inventories; (4) expenditures for new plants and equipment, book value of fixed assets, depreciation and retirements, rental payments for buildings and equipment; and (5) origin of exports of manufactured products. For all except the last category, the data are presented on the national level. (SW)
Indexed in: *ASI*

28   Industry

68. **Capacity Utilization Manufacturing and Materials.** Federal Reserve System. Board of Governors. Monthly. *Capacity Utilization in Manufacturing*
Tables. FR 1.52/2:

Provides capacity utilization (percent) for total manufacturing, primary processing and advanced processing, and manufacturing output and capacity indexes. (SW)
Indexed in: *ASI*

69. **Census of Manufactures.** Department of Commerce. Bureau of the Census. 1810–   . Quinquennial.
Tables. C 3.24/12:

Part of the economic census. Published in years ending in "2" or "7" (e.g., 1982). Since 1967, two major types of statistics are supplied: (1) general statistics, including number of establishments, employment, payroll, work hours, cost of materials, shipment values, capital expenditures, and inventories; and (2) quantity and value of materials consumed and products shipped. The data are presented in three major series: (1) Industry Series, providing detailed statistics for nearly 500 industries in each of approximately 80 industry groupings; (2) Geographic Area Series, in which a report is issued for each state and the District of Columbia detailing the manufacturing statistics for the state as a whole and also for SMSAs, counties, and incorporated places within the state; and (3) Subject Series, in which detailed reports on specific manufacturing topics are covered (e.g., Concentration Ratios). Specific topics covered change from census to census. For interim data between censuses, see the *Annual Survey* (item no. 67). (SW)
Indexed in: *ASI*

F. Natural Resources

1. Mining

70. **Census of Mineral Industries.** Department of Commerce. Bureau of the Census. 1840–   . Quinquennial.
Tables. C 3.216/[no.]:MIC[nos.]

Part of the economic census. Published in years ending in "2" or "7" (e.g., 1982). Presents a statistical picture of the mineral industries in the United States. Statistics are developed for each establishment with one employee or more, primarily engaged in mining. Aggregate

data are arranged by industry, area, employment size of establishment, and type of ownership. Some comparisons are made to earlier censuses. Three separate series make up the full volume—Subject, Geographic Area, and Industry. (WK)
Indexed in: *ASI*

71. **Minerals & Materials.** Department of the Interior. Bureau of Mines. 1976– . Bimonthly.
    Tables, graphs. I 28.149:

Covers the world mineral markets, giving production statistics, international issues, and actions on the international, federal, state, and local levels. Developments in the U.S. mineral industry are discussed in sections highlighting specific minerals. For the United States and foreign countries, two tables are given covering (1) selected new contracts, investments, expansions, and exploration activities; and (2) selected closings, postponements, layoffs, and production decreases. The Mineral Data Series gives consumption, production, imports, exports, inventories, and prices for 12 different commodities. Each issue features an article on a topic of current interest. (WK)
Indexed in: *ASI*

72. **Minerals Yearbook.** Department of the Interior. Bureau of Mines. 1882– . Annual. *Mineral Resources of the United States* (1882–1931)
    Tables, bibliographies, statistical summaries. I 28.37:

Three-volume set covering the minerals industry worldwide. Volume 1 is *Metals and Materials*, Volume 2 is *Area Reports: Domestic*, and Volume 3 is *Area Reports: International.* Chapters that will be included later in the bound volumes are issued as pre-prints in advance of publication as they are completed. Has supplement entitled *Statistical Appendix to Minerals Yearbook.* All major metallic and non-metallic mineral commodities, all 50 states and three U.S. possessions, and more than 130 foreign countries are given separate coverage. (WK)
Indexed in: *ASI*

2. Fishing

73. **Fisheries of the United States.** Department of Commerce. National Marine Fisheries Service. 1952– . Annual. *Fisheries of the United States and Alaska* (1952–58)
    Tables, graphs, glossary, publications list. C 55.309/2-2:

30   Industry

Preliminary data on the fishing industry worldwide. Final data published in *Fishery Statistics of the United States* (see item no. 74). Includes data on landings, world fisheries, production of processed fishery products, imports and exports, U.S. supply of various species, and per capital figures. Provides general administrative information for the National Marine Fisheries Service and other relevant agencies, and a list of services which can be of help to those in the fishing industry. (WK)
Indexed in: *ASI*, Self.

74. **Fishery Statistics of the United States.** Department of Commerce. National Marine Fisheries Service. 1918– . Annual.
Fishery Industries of the United States (1918–38)
Tables, glossary, photographs. C 55.316:
A statistical compendium on the fishing industry. Data are given for the United States as a whole, and broken down into ten geographic sections, such as New England, South Atlantic, Great Lakes, etc. Tables cover landings, canning, production, operating units, fishing vessels, and employment. Each section begins with a textual summary. The pictorial section provides pictures, range, and gear needed for many of the species of fish landed commercially in the United States. Publishes final data from *Fisheries of the United States* (see item no. 73). (WK)
Indexed in: *ASI*

3.   Petroleum and Natural Gas

75. **Monthly Motor Fuel Reported by States.** Department of Transportation. Federal Highway Administration. Jan. 1985– . Eight times per year (1st and 2nd month of each quarter).
Tables, charts, graphs. TD 2.46/2:
State-by-state statistics on monthly sales of gasoline, special fuels (primarily diesel fuel with small amounts of liquefied petroleum gas), and gasohol. Comparisons are made to previous years; state tax rates on motor fuel are given. This is an abbreviated version of the report *Monthly Gasoline Reported by States* (previous title: *Monthly Motor Gasoline Reported by States*) which is published quarterly. (WK)
Indexed in: *ASI*

76. **Natural Gas Monthly.** Department of Energy. Office of Oil and Gas. 1980– . Monthly.
Natural Gas Monthly Report (Sept. 1981–July 1982)
Natural and Synthetic Gas (1980–81)
Tables, graphs, maps, bibliographies, photographs. E 3.11:
Merger of: *Natural Gas Monthly Report, Underground Natural Gas Storage in the United States, U.S. Imports and Exports of Natural Gas,* and *Main Line Sales of Natural Gas to Industrial Users.* Covers the natural gas industry worldwide, providing tables that give summary statistics on producer-related activities, interstate pipeline activities, underground storage, and distribution and consumption. Also included are an industry overview, recent developments, explanatory notes, and feature articles. A similar publication is *Natural Gas Annual*, which presents a wider variety of statistics but is not as up-to-date, and focuses mainly on the United States. (WK)
Indexed in: *ASI*, Self

77. **Petroleum Marketing Monthly.** Department of Energy. Office of Oil and Gas. Apr. 1983– . Monthly.
Tables, graphs, charts, photographs. E 3.13/4:
Information about crude oils and refined petroleum products is given in seven different categories: summary statistics, crude oil prices, sales prices of petroleum products, sales volume of petroleum products, percentages of petroleum product sales, first sales of petroleum products for consumption, and standard errors and survey response percentages. Tables give detailed geographic coverage by country of origin, Petroleum Administration for Defense (PAD) districts, and individual states. Formed by the union of *Prices and Margins of No. 2 Distillate Fuel Oil* and *Monthly Petroleum Product Price Report.* (WK)
Indexed in: *ASI*

78. **Petroleum Supply Annual.** Department of Energy. Office of Oil and Gas. 1981– . Annual.
Tables, charts, maps, glossary. E 3.11/5-5:
Formed by the merger of *Crude Petroleum, Petroleum Products, and Natural Gas Liquids; Petroleum Refineries in the United States and U.S. Territories; Sales of Liquefied Petroleum Gases and Ethane in...;* and *Deliveries of Fuel Oil and Kerosene in....* Published in two volumes. Volume 1 contains two sections: (1) Petroleum Supply Summary, which gives annual data on the petroleum industry covering supply, consumption, and production; and (2) Refinery Capacity, which gives annual data on the refinery industry covering sales,

32   Industry

facilities, and capacity. Volume 2 gives statistics for each month on activities with crude oil and petroleum products. Volume 2 cumulates *Petroleum Supply Monthly* (see item no. 79). (WK)
Indexed in: *ASI*

79.   **Petroleum Supply Monthly.** Department of Energy. Office of Oil and Gas. Mar. 1982– . Monthly.
Tables, graphs, maps, glossary. E 3.11/5:

Four publications were merged to make this one: *Monthly Petroleum Statistics Report; Monthly Petroleum Statement; Supply, Disposition, and Stocks of All Oils by Petroleum Administration for Defense Districts and Imports into the United States, by Country;* and *Availability of Heavy Fuel Oils by Sulfur Level.* Provides statistics on crude oil and petroleum products by Petroleum Administration for Defense (PAD) districts. Topics include supply and disposition, production, refinery operations, imports and exports, and transportation, among others. Replaced annually by *Petroleum Supply Annual* (see item no. 78), volume 2. (WK)
Indexed in: *ASI*, Self

80.   **U.S. Crude Oil, Natural Gas, and Natural Gas Liquid Reserves.** Department of Energy. Office of Oil and Gas. 1977– . Annual. *U.S. Crude Oil and Natural Gas Reserves* (1977–81)
Tables, graphs, maps, glossary. E 3.34:

Estimates of proved reserves for crude oil, natural gas, and natural gas liquids are given for both the national and state levels. Discussions of methodologies, statistical considerations, and data collection operations are also provided. Industry developments and area highlights for such areas as the Alaska North Slope, the Gulf of Mexico, and West Texas are discussed in an overview of the industry. (WK)
Indexed in: *ASI*

81.   **Weekly Petroleum Status Report.** Department of Energy. Energy Information Administration. June 19, 1981– . Weekly. *Energy Information Administration Weekly Petroleum Status Report* (Feb. 1, 1980–June 12, 1981)
Tables, graphs, glossary. E 3.32:

Weekly statistics on the petroleum industry. Designed to be useful to the industry, consumers, and the government, it covers virtually

all kinds of petroleum products—crude oil, motor gasoline, distillate fuel oil, and residual fuel oil. Refinery activity, barrels in stock, imports, and prices are among the statistics provided. (WK)
Indexed in: *ASI*

G. Retail and Wholesale Trade

82. **Census of Retail Trade.** Department of Commerce. Bureau of the Census. 1929– . Quinquennial. *Census of Business* (1929–67)
Tables. C 3.255/

Part of the economic census. Published in years ending in "2" or "7" (e.g., 1982). This multi-volume work presents detailed statistics on all businesses engaged in selling merchandise for personal or household consumption. It is divided into three major volumes: Volume 1, Retail-Summary and Industry Statistics, includes data on establishment and firm size (including legal form of organization) for the United States; measures of value produced, capital expenditures, depreciable assets, and operating expenses for the United States; merchandise line for each state, District of Columbia, SMSAs and the United States; and miscellaneous subject reports for the United States, some individual states, and SMSAs. Volume 2, Retail Trade-Geographic Area Statistics, includes statistics for each state, SMSA, county, and places with a population of 2,500 or more. Volume 3, Retail Trade-Major Retail Centers, includes statistics for central business districts and major retail centers (shopping centers and strip malls) in each SMSA. All statistics are presented by kind of business (according to SIC codes) for each topic or geographic area. Although tables vary depending upon topic, they usually include number of establishments, sales, payroll, and number of employees. (JN)
Indexed in: *ASI*

83. **Census of Wholesale Trade.** Department of Commerce. Bureau of the Census. 1929– . Quinquennial. *Census of Business* (1929–67)
Tables. C 3.256/

Part of the economic census. Published in years ending in "2" or "7" (e.g., 1982). Contains data on wholesale establishments, with payrolls, that are primarily engaged in selling merchandise to retailers; to industrial, commercial, institutional, farm, or professional users; or to other wholesalers, excluding government organizations.

Divided into two volumes: Industry Statistics and Geographic Area Statistics. All statistics are presented by kind of business (SIC) for each topic or geographic area. Tables include number of establishments, sales, payroll, number of employees, operating expenses, inventories, and type of operation. Industry Statistics includes establishment and firm size (including legal form of organization), measures of value produced, capital expenditure, depreciable assets, and operating expenses. Commodity line sales for 15 selected SMSAs and 15 selected states, and miscellaneous subjects are also covered. (JN)
Indexed in: *ASI*

84. **Current Business Reports. BR, Monthly Retail Trade, Sales and Inventories.** Department of Commerce. Bureau of the Census. 196?– . Monthly. *Current Business Reports. BR, Monthly Retail Trade, Sales, Accounts. Receivable and Inventories.*
Tables. C 3.138/3:

This important statistical report includes estimates of sales of retail stores by kind of business (SIC) for the United States as a whole, and for geographic regions, divisions, selected states, SMSAs, Standard Consolidated Areas (SCAs), and cities. National estimates of the end-of-month inventories of retail establishments are categorized by kind of business. Data are based on samples; each issue is 35–40 pages long. Included with subscription to this publication are: *Current Business Reports, CB, Advance Monthly Retail Sales* (C 3.138/4:); *Current Business Reports, Revised Monthly Retail Sales and Inventories* (C 3.138/3-3:); and *Annual Retail Trade* (C 3.138/2-3:). Data from these publications eventually cumulate in the *Census of Retail Trade* (see item no. 82). (JN)
Indexed in: *ASI*

85. **Current Business Reports. BW, Monthly Wholesale Trade, Sales and Inventories.** Department of Commerce. Bureau of the Census. 1979– . Monthly. *Current Business Reports. Wholesale Trade Report, Sales and Inventories.*
Tables, charts, graphs. C 3.133:

This important current statistical report includes four tables: (1) estimates of monthly sales, inventories, and stock sales ratios of merchant wholesalers by kind of business (SIC codes 50 and 51) both adjusted and unadjusted for seasonal variation; (2) percentage of change; (3) coefficients of variation of sales and inventories; and (4)

seasonal adjustment factors. Also included is a chart of sales, inventories, and stock/sales ratio. Each issue is seven pages long. Statistics from this title cumulate into *Current Business Reports, Revised Monthly Wholesale Trade, Sales and Inventories* and *Current Business Reports, Wholesale Trade, Annual Sales and Year-End Inventories of Merchant Wholesalers.* (JN)
Indexed in: *ASI*

H. Service

1. General

86. **Census of Service Industries.** Department of Commerce. Bureau of the Census. 1933– . Quinquennial. *Census of Business* (1929-67)
Tables. C 3.257:

Part of the economic census. Published in years ending in "2" or "7" (e.g., 1982). A geographic series consisting of 52 separate reports makes up the first and largest section of this census. Both taxed and tax-exempt enterprises are covered, arranged by kind of business. Data are given for the United States as a whole, individual states and the District of Columbia, counties, and places with either 2,500 inhabitants or 300 or more establishments. Number of establishments, receipts, payroll, and number of employees are given in every case. Greater detail given for larger geographic entities. Comparative statistics show percent changes in selected measures since the previous census report. Appendixes explain methodology, define terms, and show specimen forms. A 5-part industry series makes up the second section of this census. The Establishments and Firm Size (Including Legal Form of Organization) volume includes income, number of employees, and number of locations for approximately 140 kinds of businesses at 2-, 3-, and 4-digit SIC levels of industry classification. The Capital Expenditures, Depreciable Assets, and Operating Expenses volume includes statistics for capital expenditures, depreciable assets, and operating expenses at the 2-digit SIC level. The remaining three reports present data for (1) hotels, motels, and lodging places; (2) the motion picture industry; and (3) other miscellaneous industry groups. Data, grouped by industry classifications, are shown for the United States as a whole and for states and SMSAs where feasible. (EW)
Indexed in: *ASI*

36   Industry

2. Health Care

87. **Health Care Financing Review.** Department of Health and Human Services. Office of Research and Demonstrations. 1979– . Quarterly.
    Tables. HE 22.18:

This publication is a forum for the discussion of issues and options for delivering cost-effective, quality health care to the elderly, poor, and disabled. It is also a place to examine the impact of ongoing health care programs on clients, professionals in the field, and the industry as a whole. Issues contain a statistical report, several research articles, a less technical "special report," and current awareness items including legislative action, administrative decisions, and grants and contracts awarded. A series of tables illustrates trends in health care financing. (EW)
Indexed in: *F&S,* INFORM, *IUSGP, PAIS*

88. **Health, United States.** Department of Health and Human Services. National Center for Health Statistics. 1975– . Annual.
    Charts, tables. HE 20.6223:

Tracking recent trends in the health care sector of the economy, this source charts, and briefly summarizes, current topics of general interest. There are detailed statistical tables in four major subject areas: health status and determinants; utilization of health resources; availability of health care; and health care expenditures. Appendixes document sources of data and serve as a glossary. (EW)

89. **Public Health Reports.** Department of Health and Human Services. Public Health Service. 1878– . Bimonthly. Title various.
    HE 20.30:

Survey and research reports focus on a wide spectrum of public health issues from very topical subjects (smokeless tobacco, AIDS, health problems of the homeless) to traditional public health concerns (family planning and food poisoning). Other regular features include brief news items on programs, practices, and people, and notification of training opportunities. (EW)
Indexed in: *ASI, IUSGP, PAIS*

I. Transportation

1. General

90. **Annual Report.** Interstate Commerce Commission. 1887– .
    Tables, charts, graphs. IC 1.1:

The source for identifying the functions and responsibilities of the ICC, it includes features about the year in review, legislation, internal administration, and energy and the environment. Considerable financial and statistical data about railroads, trucking, buses, freight forwarders, water carriers, tariffs, and court actions augment the information provided in other industry-specific publications. A directory of field offices and regional headquarters with a Commission organizational chart and a list of ICC Commissioners since 1887 enhance its use as a reference tool. The format is informal and newsy with a nice use of graphics. (LH)
Indexed in: *ASI*

91. **Census of Transportation.** Department of Commerce. Bureau of the Census. 1963– . Quinquennial.
    Tables. C 3.233/3:

Part of the economic census. Published in years ending in "2" or "7" (e.g., 1982) but the 1963 census was the first of this type. There are four major projects: the Commodity Transport Survey, the Passenger Transport Survey, the Truck Inventory and Use Survey, and the Bus and Truck Carrier Survey. The 1982 Truck Inventory and Use Survey is issued in 52 separate parts—one for each state—and compared with previous surveys. Summaries of the number of trucks and related detailed characteristics, such as mileage and type of freight transported, are given. (LH)
Indexed in: *ASI*

92. **National Transportation Statistics.** Department of Transportation. Research and Special Program Administration. 1977– . Annual.
    Tables, charts, graphs, bibliography, glossary. TD 10.9:

Supports the mission of the DOT's Transportation Systems Center to disseminate national transportation energy statistics. The major advantage of this publication is that ten modes of transportation are profiled, compared over a period of ten years, and gathered into one place. All statistics are derived from government agencies and trade associations and, furthermore, all sources of the data are documented. Volume 1, 1977, covered the 11 calendar years ending in

38   Industry

1975. Tree displays illustrating the interrelations of transportation modes are also provided. (LH)
Indexed in: *ASI*

93.   **Transport Economics.** Interstate Commerce Commission. Bureau of Economics. 1941–   . Monthly. *Monthly Comment on Transportation Statistics* (Aug. 1941–Apr. 1955)
Tables. IC 1.17:

Primary purpose is to present current statistics and other material not readily available from other sources in a convenient and accessible format. Statistics are cumulated and included in the *Transport Statistics in the United States* (see item no. 94). (LH)
Indexed in: *ASI*

94.   **Transport Statistics in the United States.** Interstate Commerce Commission. Bureau of Accounts. 1954–   . Annual. Supersedes the *Annual Report on the Statistics of Railways in the U.S.*, 1st to the 87th Annual Reports (1888–1953)
Tables. IC 1.25:

Released in nine separate parts, issued individually, as the data are collected, compiled, and published. The parts are: (1-3) Railroads, their lessors, and proprietary companies; (4) Electric Railways; (5) Carriers by Water; (6) Oil Pipe Lines; (7) Motor Carriers; (8) Freight Forwarders; and (9) Private Car Lines. Preceded by individual issues of *Transport Economics* (see item no. 93), that are gathered to create this publication. (LH)
Indexed in: *ASI*

2.   Airlines

95.   **Air Carrier Financial Statistics.** Department of Transportation. Office of Aviation Information. 1962–   . Quarterly.
Tables. C 31.240:

An excellent source for current, in-depth information. Arranged by individual airline, it presents the operating revenues, expenses, and profits for the major national and large regional airlines. Contains a glossary of terms and technical notes, as well as semiannual tables. The Civil Aeronautics Board (CAB), abolished in early 1985, issued this publication until June of 1984 when it was transferred to the Department of Transportation. It is fortunate that the DOT has decided to maintain this publication as it is an extremely vital source in the field. It is based on the CAB Form 41 financial report

that is completed by each airline. *ASI* says this form is still being utilized. (LH)
Indexed in: *ASI*

96. **Air Carrier Traffic Statistics.** Department of Transportation. Office of Aviation Information. 1962– . Monthly. *Monthly Report of Air Carrier Traffic Statistics* (Jan. 1956–Jan. 1962) Tables. C 31.241:

As with the quarterly publication, *Air Carrier Financial Statistics* (see item no. 95), this was also published by the CAB until August, 1984, when the responsibility for it was transferred to the DOT. Even more current and vital than its quarterly counterpart, its tables show changes in revenue load factors, revenue ton miles, revenue passenger miles, average revenue load and passenger load, average seats available, seat-mile capacity, schedule performance factors, and revenue aircraft miles. In addition, it provides 2-year comparisons of the above figures. Extremely pertinent information for the travel industry. (LH)
Indexed in: *ASI*

97. **FAA Statistical Handbook of Aviation.** Department of Transportation. Federal Aviation Administration. 1944– . Annual.
    CAA Statistical Handbook of Aviation (1958–68)
    Statistical Handbook of Civil Aviation (1944–57)
    Tables, charts, graph, maps. TD 4.20:

A general handbook of statistics on the FAA, presented in tabular form. Included are airports, airways, airmen certification, air passengers and commuters, general aviation, aircraft, aeronautical production and exports, scheduled air carrier operations, and accidents. A convenient source for historical data, its primary purpose is stated as "to assist in evaluating progress." Historical tables, a glossary of terms and acronyms, a map of FAA regional boundaries, and an index make this publication indispensable to all collections with an interest in aviation. (LH)
Indexed in: *ASI*

98. **Handbook of Airline Statistics.** Department of Transportation. Research and Special Programs Administration. 1938– . Annual.
    Title varies: 1938–48; *Annual Airline Statistics*; or *Handbook of Airline Statistics, United States Certified Air Carriers.*
    Tables, maps. C 31.249:

40   Industry

A requisite reference source for documenting the state-of-the-art of the commercial air transportation industry in the United States. Permits identification of industry trends and a detailed look at all aspects of the financial situation of the airline industry. Updated monthly by *Air Carrier Traffic Statistics* (see item no. 96), and quarterly by *Air Carrier Financial Statistics* (see item no. 95). (LH)
Indexed in: *ASI*

99. **U.S. International Air Travel Statistics.** Department of Transportation. Transportation System Center. 1979– . Monthly with quarterly and annual summaries.
    Tables, charts, graphs. TD 1.40/3:
A straight statistical, no frills publication that provides ready information on commercial air passenger travel between the United States and foreign countries. The Immigration and Naturalization Service collects the data. U.S. arrivals and departures of alien citizens are noted. The inclusion of 5-year travel trends enhance the usefulness of this publication for the travel industry and other related industries. (LH)
Indexed in: *ASI*

3. Highways

100. **Highway Statistics.** Department of Transportation. Federal Highway Administration. 1945–   . Annual.
    Tables, maps. TD 2.23:
An extensive publication presenting statistics on motor fuel and vehicles, driver licensing, highway-user taxation, state highway finance, highway mileage, and federal aid for highways. Includes U.S. territories, Samoa, Guam, Puerto Rico, and the Northern Marianas. Much of the information presented in the earlier editions is summarized in the publication *Highway Statistics Summary to 1985* (TD 2.23/2:); in 1986 a 40th anniversary edition was published covering the years 1945–85. Identifies and consolidates data useful for every level of society. (LH)
Indexed in: *ASI*

101. **Public Roads, A Journal of Highway Research and Development.** Department of Transportation. Federal Highway Administration. 1918–   . Quarterly.
    Charts, graphs. TD 2.19:
A technical periodical designed for those actually engaged in planning and constructing highways, and for instructors of highway

engineering. This is the source to consult for identifying new research in progress. The title, Federal Cataloging Program category and numbers, research objectives, performing organization, expected completion date, and estimated cost are given. Photographs enhance the usefulness of the articles, which are well-documented with extensive bibliographies. Excellent indexing adds to its reputation as a reliable, accurate, and timely publication. (LH)
Indexed in: *ASI, IUSGP*

4. Maritime

102. **Maritime Subsidies.** Department of Commerce. Office of Trade Studies and Statistics. 1958– . Biennial.
Tables. TD 11.27:

This publication's prime purpose is to present subsidization information, and the inclusion of the economic background of each nation and significant events in its maritime industries make this a good source for historical information as well. Being able to identify the countries that are most heavily subsidized in this area is very useful for industrial decision making. Since this information is presented comparatively by nation, it affords its users a handy way to view the overall picture. (LH)

103. **Merchant Fleets of the World: Oceangoing, Steam, and Motor Ships of 1,000 Gross Tons and Over, as of Jan. 1, 1984.** Department of Transportation. Studies and Subsidy Contracts. 1955– . Annual. *Merchant Fleets of the World* (1955–84)
Tables, charts, graphs. TD 11.14:

Gives the number and tonnage of ships in world merchant fleets and their yearly changes—additions or deletions. The World Summary has one table for each of the 14 largest fleets and is divided into four types of ships: freighters, bulk carriers, tankers, and combination passenger/cargo. The total ships in each fleet, new ships by type and country, and ships lost or swapped are included. A necessary reference source for any company engaged in waterborne commerce, it appears regularly. (LH)
Indexed in *ASI*

104. **New Ship Construction: Oceangoing Ships of 1,000 Gross Tons and Over in U.S. and Foreign Shipyards.** Department of Transportation. Maritime Administration. 1946– . Annual.

42   Industry

Merchant Ships Built in the United States and Other Countries: Seagoing Steam and Motor Ships of 1,000 Gross Tons and Over (1946-62)
Tables. TD 11.15:
Gives an idea of the state of the ship construction industry throughout the world. It covers the leading 20 countries, their deliveries and orders. Four types of merchant ships are arranged by country of construction and country of intended registration with excellent cross-tabulations. (LH)
Indexed in: *ASI*

105.   **Waterborne Commerce of the United States.** Department of the Army. Corps of Engineers. 1942-   . Annual.
Tables. D 103.1/2:
Since 1953, this report has been issued in five parts, bound separately, one for each region of the country. It was formerly part two of *The Annual Report* of the Chief of Engineers of the U.S Army. The data are based on vessel operator reports, as the Army Corps of Engineers has been responsible for data collection in this area since 1866. Parts 1-4 cover Waterways and Harbors: Part 1, Atlantic Coast; Part 2, Gulf Coast, Mississippi River System, and Antilles; Part 3, Great Lakes; Part 4, Pacific Coast, Alaska, and Hawaii. Part 5 contains national summaries of the regional statistics. A timely and "on time" publication. (LH)
Indexed in: *ASI*

5.   Railroads

106.   **Financial and Operating Statistics: Class I Railroads.** Interstate Commerce Commission. Bureau of Accounts. 1972-   . Semiannual.
Operating Revenues and Operating Expenses of Class I Railroads in the U.S. (1920-71)
Operating Statistics of Large Railroads, Selected Items (1920-71)
Revenue Traffic Statistics of Class I Railroads in the U.S. (1921-71)
Selected Income and Balance Sheet Items of Class I Railroads in the U.S. (1932-71)
Motive Power and Car Equipment of Class I Railroads in the U.S. (1936-71)

Train and Yard Service of Class I Railroads in the U.S. (1964-71)
Tables. IC 1 ste.19/4:

The titles of the publications that were subsumed clearly indicate the contents of this all-in-one necessary source for railroad research, be it historical or current. No other publications exist that document in such detail and with such preciseness the nature of the railroad industry in the United States. (LH)
Indexed in: *ASI*

107. **Freight Commodity Statistics of Class I Railroads in the U.S.** Interstate Commerce Commission. Bureau of Accounts. 1920– . Quarterly.
Tables. IC 1 ste.26:

Gives tons of revenue carried and total freight revenue, freight traffic originated and terminated, total freight carried, and freight revenue by class of commodities and by regions. The only source to consult for reliable information in this area, it is supplemented each year (since 1924) by a similar publication, *Freight Commodity Statistics of Class I Railroads in the U.S.*, an annual summary of these quarterly publications (IC 1 ste.29). (LH)
Indexed in: *ASI*

108. **Rail Carload Cost Scale by Territory.** Interstate Commerce Commission. Bureau of Accounts. 1948– . Annual.
Tables. IC 1 acco.7:

The sole purpose of this publication is to allow users to calculate the cost of individual shipments in single car, multiple car, or unit train service by applying costing methodology to specific shipments. Principal tables include carload mileage cost scales by district and region and by type of car; and carload unit costs for various weight loads by type of equipment. A useful source used extensively by those in the industry. (LH)
Indexed in: *ASI*

J. Travel

109. **Forecasts of Citizen Departures to Europe for [date].** Department of State. Bureau of Consular Affairs, Passport Services. 1980?– . Monthly.
Tables. S 1.121:

44   Industry

A single sheet published monthly giving passport officer estimates on the number of U.S. citizens who will be traveling to Europe the following month. Tables list volume of passport applications and citizen departures to Europe for the preceding 13 months and the percentage of change in each case over the previous year. (JN)
Indexed in: *ASI*

110. **Summary and Analysis of International Travel to the U.S.** Department of Commerce. United States Travel Service. 1974– (suspended 1979–85). Quarterly.
Tables. C47.15:

Presents highlights of foreign tourist arrivals to the United States for the current quarter. Statistical tables break down the information into arrivals by country of origin, geographic area, mode of transport, and the region of permanent residence. Current statistics are presented with those from the preceding year and a percent of change is indicated. (JN)
Indexed in: *ASI*

K.   Utilities

111. **Annual Energy Review.** Department of Energy. Energy Information Administration. 1982– . Annual.
Tables, charts, graphs, maps. E 3.1/2:

Brief discussion of major energy developments of the year is followed by historical data (1949 to present) which adds to the information in many sections of *Monthly Energy Review* (see item no. 117). Sections are devoted to petroleum, natural gas, coal, electricity, nuclear energy, wood, waste, solar and geothermal energy. Production, consumption, imports and exports, prices, and storage are a few of the broad categories of statistical data included. Specific consumption tables are very detailed, e.g., energy use for various household appliances. A final section covers international energy resources, focusing on supply and disposition. Before 1982, this publication was volume two of the EIA *Annual Report*. (RB)
Indexed in: *ASI*

112. **Electric Power Annual.** Department of Energy. Energy Information Administration. 1982– . Annual.
Electric Energy and Peak Load Data—Annual [1981]
Power Production, Fuel Consumption, and Installed Capacity Data—Annual Report (1980)
Tables. E 3.11/17-10:

Summarizes electric utility industry statistics on an annual basis for the last five years at the national, regional, and state level. The standard areas of generation, consumption, receipts, and stocks of major energy sources are detailed. Retail prices and sales of electricity receive more coverage than in *Electric Power Quarterly* (see item no. 114) and *Electric Power Monthly* (see item no. 113). Comparison table for 40 selected cities also covers five years. (RB)
Indexed in: *ASI*

113.   **Electric Power Monthly.** Department of Energy. Energy Information Administration. 1980–   . Monthly.
       Preliminary Power Production, Fuel Consumption and Installed Capacity Data
       Energy and Peak Load Data
       National Trends in Electric Energy Resources
       Residential Electric Bills in Major Cities
       Tables. E 3.11/17-8:
Electric utility industry statistics on a census division and state level. Monthly data and year-to-date comparisons are available for all major energy sources, focusing on generation, consumption, and end-of-month stocks. Yearly totals for electricity sales in kilowatt hours for residential, commercial, and industrial service are given from 1976 to the present. Comparison of retail prices for electricity in 40 selected cities provides a very abbreviated version of *Typical Electric Bills* (see item no. 119). (RB)
Indexed in: *ASI*

114.   **Electric Power Quarterly.** Department of Energy. Energy Information Administration. 1983–   . Quarterly. Continues in part: *Cost and Quality of Fuels for Electric Utility Plants*
       Tables. E 3.11/17-11:
Detailed electric utility statistics include monthly summaries and quarterly totals for the reporting quarter, from national to plant level. Coal, petroleum, and gas receipts, costs, and quality figures are given at the census division and state level by plant size. The primary section, Plant Statistics, lists plants and provides net generation data by type of fuel, fuel consumption, and fuel stocks. Includes quarterly summary of reported major disturbances and unusual occurrences at electric utilities. (RB)
Indexed in: *ASI*

46   Industry

115. **Financial Statistics of Selected Electric Utilities.** Department of Energy. Energy Information Administration. 1982–   . Annual.

Tables, charts, graphs. E 3.18/4:

Presents extensive data on electric power generation and financial operations of major electric utilities, privately and publicly owned. Privately owned company statements are arranged by state, permitting easy comparison of balance sheets, revenues, sales, numbers of customers, expenses, utility plants, and physical quantities. Financial and operating data for publicly owned electric utilities are presented in composite tables covering the last five years. Brief appendixes include similar summary data for Rural Electric Cooperative Borrowers and Federal Power Marketing Agents. Combines: *Statistics of Privately-Owned Electric Utilities* (1966–81) and *Statistics of Publicly-Owned Electric Utilities* (1966–81). Prior to 1966: *Statistics of Electric Utilities in the U.S., Classes A & B, Privately-Owned Companies* and *Statistics of Electric Utilities in the U.S., Classes A & B, Publicly-Owned Companies.* (RB)

Indexed in: *ASI*

116. **Inventory of Power Plants in the United States.** Department of Energy. Energy Information Administration. 1975–   . Annual.

Tables, charts, graphs, maps. E 3.29:

Reports on electric generating units, operating and projected, in the 50 states, District of Columbia, and Puerto Rico. Tables present aggregate data on the national and regional levels for operable and planned units by energy source, showing number of units and megawatt capability. The major section profiles operable power plants by state, company, and plant. For each plant, capability in megawatts, type of unit, energy source, and year of initial operation is given. Projected units for the next ten years are listed by state, company, and plant with completion data, capacity/capability, unit type, energy source, and current status. Published semiannually 1975–77. (RB)

Indexed in: *ASI*

117. **Monthly Energy Review.** Department of Energy. Energy Information Administration. 1974–   . Monthly.

Tables, charts, graphs. E 3.9:

Statistical report on the domestic production, consumption, prices, stock, imports and exports of petroleum, natural gas, coal, electricity, and nuclear energy. Tables include annual average data from

1973 to the present, with monthly figures for the last three years. International section includes production for major petroleum producing countries, consumption and stocks for OECD countries, and nuclear electricity generation for non-communist countries. Conversion factors provide the approximate heat content/rates of energy sources. (RB)
Indexed in: *IUSGP*

118. **Short-Term Energy Outlook.** Department of Energy. Energy Information Administration. 1979– . Quarterly.
Tables, charts, graphs. E 3.31:

Narrative report and tables provide forecasts of international and domestic energy supply, demand, and prices on a quarterly basis for the current year, with annual figures for the coming year. Primary emphasis is given to petroleum products, especially gasoline and fuel oil, but natural gas, coal, and electric power are included. Tabular projections are supported by concise, readable discussions of macroeconomic developments, seasonal influences, and domestic trends. (RB)
Indexed in: *ASI*

119. **Typical Electric Bills.** Department of Energy. Energy Information Administration. 1959– . Annual.
Tables, maps. E 3.22:

Detailed information on typical electric bills for residential, commercial, and industrial service. Including communities of over 2,500 inhabitants in all 50 states, District of Columbia, and Puerto Rico, residential data are arranged by five levels of kilowatt-hour consumption, and by broad types of use, e.g., space heating. The Commercial and the Industrial sections include cities of over 50,000 inhabitants, and show billing demand and consumption in kilowatt hours. All sections name the utility serving the community. Several tables rank communities by consumption and type. Appendix provides state averages for rural cooperatives, public and private utilities. Combines: *Typical Electric Bills, Cities of 50,000 Population and More* and *Typical Residential Electric Bills.* (RB)
Indexed in: *ASI*

VI. AGRICULTURE

120. **Agricultural Outlook.** Department of Agriculture. Economic Research Service. June 1975– . Monthly (except combined Jan./Feb. issue).
Tables, graphs. A 93.10/2:

Covers the agriculture industry of the United States in articles and statistical tables. All varieties of livestock and crops are included. Articles give prospects for coming months, explanations of recent events, and analyses of trends in various segments of the industry. Tables cover agricultural economic indicators, market prices, supply and utilization of products, and imports and exports. (WK)
Indexed in: *ASI, IUSGP, PAIS*

121. **Agricultural Prices.** Department of Agriculture. Crop Reporting Board. June 1942– . Monthly. *Farm Product Prices*
Tables, graphs. A 92.16:

Volumes for 1980–82 are summarized by *Agricultural Prices, Annual Summary*; 1983–present summarized by *Agricultural Prices...Summary*. Each issue has two main parts: Prices Received by Farmers and Prices Paid by Farmers. Prices for each crop and type of livestock are given and compared with earlier years. Tables are arranged by state or by type of product. Trends in agricultural price indexes are analyzed. (WK)
Indexed in: *ASI*

122. **Agricultural Statistics.** Department of Agriculture. 1936– . Annual.
Tables. A 1.47:

Before 1936 the information contained in the present volume was published in the statistical section of the *Yearbook of Agriculture*

Agriculture   49

(see item no. 131). This is a compendium of the year's statistics for agriculture in the United States. Included are data on agricultural production, supplies, consumption, facilities, prices, and foreign trade. Tables giving weights, measures, and conversion factors are supplied to aid in interpreting the data. Many of the tables give historical data going back 15 years. (WK)
Indexed in: *ASI*, Self.

123.   **Census of Agriculture.** Department of Commerce. Bureau of the Census. 1840-   . Quinquennial.
       Tables, charts, maps. C 3.31/4:

Provides a statistical picture of agriculture in the United States on the national, state, and local levels. Separate volumes are issued for each state (including the District of Columbia), a United States summary, Guam, Puerto Rico, and the Virgin Islands. The state volumes contain county data as well as aggregate data for the state. All farming, ranching, and related activities are included. Each census is compared with the previous census. Beginning with 1982, the census has been taken covering years ending in "2" and "7". (WK)
Indexed in: *ASI*

124.   **Crop Production.** Department of Agriculture. Crop Reporting Board. 1941-   . Monthly.
       Tables, maps. A 92.24:

Some issues contain *Crop Production, Tobacco,* which is issued separately. Each issue is summarized in *Crop Production, Annual Summary. Crop Production* also absorbed *Farm Numbers.* Production, yield, area planted, and area harvested are given for numerous major crops, including fruits, nuts, potatoes, tobacco, wheat, and rye. The progress of each crop is discussed and a forecast is made of future production. (WK)
Indexed in: *ASI*

125.   **Economic Indicators of the Farm Sector.** Department of Agriculture. Economic Research Service. 1979-   . Annual.
       Tables. A 93.45:3

Contains five different series, each of which is published as a separate subtitled issue. Series titles are: Costs of Production; Farm Sector Review; Production and Efficiency Statistics; National Financial Summary (formerly Income and Balance Sheet Statistics); and State Financial Summary (formerly State Income and Balance Sheet Statistics). Each series gives statistical information on its particular subject along with articles discussing the data in the tables. Data are

50   Agriculture

primarily financial-cash receipts, expenses, loans, debts outstanding, production costs, labor, etc. Statistics were formerly published in *USDA Statistical Bulletin* (1923-78). (WK)
Indexed in: *ASI*

126. **Foreign Agricultural Trade of the United States.** Department of Agriculture. Economic Research Service. Aug. 1962-   . Bimonthly.
Tables, graphs. A 93.17/7:

Was published monthly, with some issues combined, from August 1962 to December 1979. Another title is *Foreign Agricultural Trade.* Supplemented by *U.S. Foreign Agricultural Trade Statistical Report* and *U.S. Foreign Agricultural Trade Statistical Report Supplementary Tables.* Formed by the union of *Foreign Agricultural Trade of the United States Digest* and *Foreign Agricultural Trade of the United States, Statistical Report.* Provides comprehensive data on quantity, value, and unit value of U.S. exports and imports. Data are arranged by commodity or commodity and country. Short articles throughout provide an overview and analysis of the industry. (WK)
Indexed in: *ASI*

127. **Foreign Agriculture.** Department of Agriculture. Foreign Agricultural Service. Jan. 7, 1963-   . Monthly.
Tables, graphs, photographs. A 67.7/2:

Published weekly from January 7, 1963 to March 19, 1979. Formed by the merger of an earlier publication with the same title and *Foreign Crops and Markets.* Each issue contains feature articles which discuss agriculture in other countries, including such topics as markets, production, and the agricultural economy. U.S. exports to certain countries are also frequent topics. Monthly departments include Marketing News, Fact File, and Country Briefs. (WK)
Indexed in: *ASI, IUSGP, PAIS*

128. **Marketing Research Reports.** Department of Agriculture. Apr. 1952-   . Irregular.
A 1.82:

Each issue, produced by a different agency, presents a report about some aspect of the agriculture industry. Topics are extremely varied, but typical reports cover such subjects as planning food distribution facilities, improving methods for packing broilers, the egg products industry, and costs of hauling bulk milk. Special features vary with each issue but explanatory diagrams, photographs, and statistics are often included. (WK)

Agriculture 51

129. **World Agricultural Supply and Demand Estimates.** Department of Agriculture. Economics and Statistics Service [and] Foreign Agricultural Service. 1973?- . Monthly, with 4 additional issues per year (Jan., Apr., June, Oct.). *Agricultural Supply and Demand Estimates* (Dec. 1973?-Sept. 1980). Tables. A 93.29/3:

Statistics are given on stocks, production, imports, exports, and consumption for livestock and crops in the United States and foreign countries. Figures for previous years are listed and projections for the coming year are made. Highlights of the projections are discussed in an introduction. (WK)
Indexed in: *ASI*

130. **World Agriculture Situation and Outlook Report.** Department of Agriculture. Economic Research Service. 1962- . Quarterly.
World Agriculture Outlook and Situation Report (Dec. 1983-Mar. 1986)
World Agriculture Outlook and Situation (1981-83)
World Agricultural Situation (1962-81)
Tables, graphs. A 93.29/2:

Textual information, supplemented by tables, analyzes the state of the world agriculture industry. Prices, consumption, production, and other developments are discussed with the aim of providing an understanding of the industry as it is now and where it will be in the future. Developments within specific countries, regions, and specific commodities are reviewed. (WK)
Indexed in: *ASI, IUSGP*

131. **Yearbook of Agriculture.** Department of Agriculture. 1890- . Annual.
Agriculture Yearbook (1923-25)
United States Department of Agriculture Yearbook (1920-22)
Yearbook of the United States Department of Agriculture (1894-1919)
Report of the Secretary of Agriculture (1890-93?)
Photographs, diagrams. A 1.10:

Before 1936, each issue contained the statistics which are now published in *Agricultural Statistics* (see item no. 122), as well as numerous articles about new developments in agriculture. Beginning in 1936, *Yearbook of Agriculture* was devoted to discussion of a single subject, while the statistical tables appeared in *Agricultural Statistics*. The discussion in the *Yearbook* is accomplished through the

use of many articles. Examples of typical volume titles are: *Better Plants and Animals, That We May Eat* (focusing on State Agricultural Experiment Stations), *Cutting Energy Costs*, and *Research for Tomorrow*. (WK)
Indexed in: Self

VII. ENVIRONMENT

132. **EPA Journal.** Environmental Protection Agency. Office of Public Awareness. 1975– . Monthly.
    EP 1.67:

Each issue covers a different topic, recent ones being hazardous waste, acid rain, and water pollution. The articles, written by EPA staff members, often cover topics relating to changes in EPA regulations. Useful for general information on regulations and potential areas of concern. (PB)
Indexed in: *IUSGP, PAIS*

133. **EPA Publications Bibliography.** Environmental Protection Agency. Library Systems Branch. 1975– . Quarterly. *EPA Reports Bibliography* (1975–76)
    EP 1.21/7-2:

Lists, indexes, and abstracts EPA technical reports and journal articles added to the NTIS collection. Indexed by title, key word, sponsoring EPA office, corporate author, personal author, contract/grant number, and NTIS order/report number. The fourth issue of every year contains cumulative indexes for the entire previous year. Continues *EPA Cumulative Bibliography,* published 1970–76. Supplemented by *EPA Publications Bibliography* (1977–83) and *Quarterly Abstract Bulletin* (1975–). Available from NTIS. (PB)

VIII. LABOR

134. **Annual Report.** Equal Employment Opportunity Commission. 1965/66– . Annual.
    Tables. Y 3.Eq2:1/

The annual report details EEOC operations and activities to eliminate job discrimination on the basis of race, color, religion, sex, national origin, age, or handicap. The textual information outlines enforcement through litigation, compliance activities, and advice, guidance, and research services. Tables provide data for the number of discrimination charges based on EEOC computer records. (WT)

135. **Area Wage Survey.** Department of Labor. Office of Wages and Industrial Relations. Annual.
    Tables. L 2.3/2:

The survey series reports occupational earnings and supplementary benefits for office and plant occupations. It outlines data in six broad industry divisions for some 90 labor market areas with more detailed reports available for approximately 90 MSAs and PMSAs. The data include establishments in manufacturing, public utilities, wholesale trade, retail trade, finance, and service industries. The earnings data will be collected and updated annually for the 32 larger areas, and the remaining 58 areas collected and updated every three years for the 32 larger areas and the remaining 58 areas collected and updated on a four-year cycle. (WT)

136. **Dictionary of Occupational Titles.** Department of Labor. Employment Service. 1939– . Irregular.
    L 37.302-OC1

The dictionary lists and defines approximately 20,000 job titles, providing comprehensive occupational information gathered during an extensive research and verification project which included on-site observations. The 1982 supplement contains occupations which have emerged since the fourth edition (published in 1977) or which were inadvertently excluded from this edition. The occupational definitions list the occupational code number and title, industry designation, alternative titles, and any undefined related titles. (WT)

137. **Employment and Earnings.** Department of Labor. Bureau of Labor Statistics. 1942- . Monthly.
Employment and Earnings and Monthly Report on the Labor Force (1966-69)
Monthly Report on the Labor Force (1959-65)
Current Population Reports P-57, Labor Force (1947-59)
Monthly Report on the Labor Force (1942-47)
Estimates of Labor Force, Employment, and Unemployment in the United States During the Week of... (1942)
Tables. L 2.41/2:

Organizing data from the Current Population Survey of household interviews and Current Employment Survey of payroll reports, this summary gives statistics for employment, unemployment, hours, and earnings. Monthly and quarterly tables detail and present further statistics which complement the coverage in the *Monthly Labor Review* (see item no. 140). Details include payrolls by 2-, 3-, and 4-digit SIC codes, and payrolls by state and selected metropolitan areas. Recurring features and articles, along with explanatory notes, complete this summary report. An annual supplement cumulates and duplicates some 25 tables from the monthly summary reports. (WT)
Indexed in: *F&S, IUSGP, PAIS*

138. **Government Employment.** Department of Commerce. Bureau of the Census. Annual.
Tables. C 3.140/2:

This series consists of four volumes: *City Employment in...* (-3: 1944-), *County Employment in...* (-5: 1974-), *Local Government Employment in Major County Areas* (-6: 1970-), and *Public Employment in...* (-4: 1948-). Each volume compares annual full- and part-time employment and payroll data for governmental functions. Data gathered from mail questionnaires summarize federal, state, county, and local government employment into a textual overview and tables. (WT)

56   Labor

139. **Handbook of Labor Statistics.** Department of Labor. Bureau of Labor Statistics. 1926–   . Irregular.
L 2.3/5:

As a compilation of the major BLS statistical series, this handbook provides data on labor conditions and labor force characteristics. Data sections cover employment, unemployment, employee characteristics, employees on non-agricultural payrolls, productivity, compensation, prices, unions, and selective statistics on foreign labor. Issued as part of the *BLS Bulletin* series, it includes historical data when available and reliable, but may differ from previous editions due to revisions. Each section usually contains technical notes to outline the programs and sources for that section. (WT)

140. **Monthly Labor Review (MLR).** Department of Labor. Bureau of Labor Statistics. 1915–   . Monthly. *Monthly Review of the U.S. Bureau of Labor Statistics* (1915–18)
Tables, book reviews. L 2.6:

This journal reports labor conditions with statistics, regular features, and articles. The current labor statistics are made up of some 48 tables and list employment, compensation, price, productivity, international comparisons, and occupational injury and illness data. Regular features include developments in industrial relations, labor month in review, expiring major labor arguments, and other summary research reports. The articles further assess labor and working conditions and usually contain statistical data. *Annual Statistical Supplement* published from 1959 to 1965. (WT)

Indexed in: *BI*, *BPI*, *F&S*, INFORM, *IUSGP*, MC, *PAIS*, *RG*, T&I, Self

141. **Occupational Outlook Handbook.** Department of Labor. Bureau of Labor Statistics. 1949–   . Biennial.
L 2.3/4:

Providing information for prospective employment, this handbook contains job expectations and conditions for approximately 200 occupations. Information includes data from BLS surveys as well as career information for those seeking vocational guidance. Supplemented by the *Occupational Outlook Quarterly* (L 2.70/4:) which selectively updates the information in the Handbook. The Quarterly was first published in 1957 and appeared under the title *Occupational Outlook: Current Supplement to Occupational Outlook Handbook* through 1958. (WT)

Indexed in: *BI*, INFORM, *IUSGP*, T&I

142. **Personnel Literature.** Office of Personnel Management. Library. 1941– . Monthly.
PM 1.16:

This index provides access to books, journal articles, and other documents received at the OPM Library. Arranged by general personnel management subjects, the citations include subject descriptors which may be used to further identify other materials in the annual subject index. The annual name index includes personal and corporate authors as well as other contributors responsible for the listed items. (WT)

143. **Standard Occupational Classification Manual.** Department of Commerce. Office of Federal Statistical Policy and Standards. 1977– . Irregular.
C 1.8/3-Oc1/980 (2nd ed.)

Created to compile statistics on the labor force, the manual contains all occupations performed for a salary or for a profit. Although some nonprofit positions appear in the classification, those positions unique to volunteer organizations were not included. The classification system numerically organizes occupations into a 4-level system: division, major group, minor group, and unit group. The system facilitates the evaluation process for employment, wages, and other data as well as providing comparison relationships between occupations and demographic characteristics. Each title includes a cross-reference number to the 4th edition of the *Dictionary of Occupational Titles* (see item no. 136). (WT)

144. **Trade and Employment.** Department of Commerce. Bureau of the Census and Department of Labor. Bureau of Labor Statistics. 1984– . Quarterly.
Tables, graphs. C 3.269:

This report measures variations in U.S. imports and related domestic employment. The comparison data regroup commodities and employment by SIC codes. Data analysis covers imports for consumption, import commodity groups with increase of 25% or more and $10 million or more over year-ago period, groups with less than 25% and $50 million or more over year-ago period, non-agricultural employment by division and mining and manufacturing industries, and employment change in selected mining and manufacturing industries. The first report, issued in November 1984, includes data for the second quarter of 1982 through the second quarter of 1984. (WT)

58   Labor

145.   **Unemployment in States and Local Areas.** Department of Labor. Bureau of Labor Statistics. 1976–   . Monthly.
Employment and Unemployment in States and Local Areas (1981–Mar. 1984)
State, County, and Selected City Employment and Unemployment (1976–80)
L 2.41/9-

This service gives monthly estimates for labor market areas, counties, and cities with a population of 25,000 or more. Arranged by states, the estimates summarize the number in the labor force, the number employed, and the number and percentage unemployed. An irregularly published supplement revises figures for the cumulated annual. (WT)

IX. SMALL BUSINESS

146. **Annual Report.** Small Business Administration. 1953/54– .
Annual.
Tables, graphs. SBA 1.1:
Two-volume set providing a general overview of the small business climate in the past year. Several tables provide statistics on business starts, failures, and bankruptcies; the number of small businesses categorized by industry; the number of people employed by small businesses; and several other types of data. This is a useful source for finding a summary of the economic conditions related to small business, but it does not provide as much statistical data as the *State of Small Business* (see item no. 153). Published semiannually 1954–61. (VS)
Indexed in: *ASI*

147. **Business Basics.** Small Business Administration.
Tables, graphs, charts. SBA 1.19:
Series of self-instructional booklets of varying lengths covering fundamentals of small business management. Each booklet includes explanatory material and exercises covering the topic being examined. An instructor's guide and summary is also provided so that the series can be used in a classroom situation. (VS)

148. **Franchise Opportunities Handbook.** International Trade Administration and Minority Business Development Agency. 1965– . Annual. *Franchise Company Data for Equal Opportunity in Business* (1965–70)
Bibliography. C 61.31:

60   Small Business

Directory listing franchisors and giving a summary of the terms required to obtain a franchise. A brief introduction explains what franchising is, offers precautionary advice on investing in a franchise, and includes a checklist for evaluating franchise opportunities. Brief entries describe the franchise operation, equity capital required, financial training, and managerial assistance provided by the franchisor. Information is arranged by type of franchise with an index in alphabetical order by franchise name. (VS)

149. **Franchising in the Economy.** Department of Commerce. Bureau of Industrial Economics. Annual.
Tables, charts, graphs. C 61.31/2:

This series contains the results of an annual survey covering two years, mailed to all types of franchisors. The material provided is heavily statistical and includes data on types of franchises, number of establishments, types of ownership, sales figures, and sales of products and services from franchisors to franchisees. Minority ownership rankings of types of franchises by sales and number of establishments, and detailed statistics on types of franchises and their operations, are also given. Although somewhat narrow in scope, a valuable source of data on franchises. (VS)
Indexed in: *ASI*

150. **Small Business Management Series.** Small Business Administration. 1952– . Irregular.
Bibliography, tables, charts, graphs. SBA 1.12:

Pamphlets of approximately 50 pages each offering advice and information on general management topics. Each booklet contains several chapters which lead the reader through a conceptual discussion of the topic and then provide instructions on how to apply this information to real-life situations. (VS)

151. **Small Business Subcontracting Directory.** Small Business Administration. 1960?– . Annual.
SBA 1.2/10:

Directory of U.S. government prime contractors that offer the most opportunities for subcontracting to small business. The prime contractors are monitored by the SBA and are listed alphabetically by name within SBA region. Each company entry includes name, division, address, small business representative including title and phone number, and type of business. Unfortunately, indexes of companies listed alphabetically or by type of business are not included. (VS)

152. **Starting and Managing Series.** Small Business Administration. 1958– . Irregular.
Bibliography. SBA 1.15:

Booklets containing fairly detailed information about general management topics and how to start specific types of businesses. These include practical advice and easily understood explanations of the topics covered. This series would be useful for people contemplating a new business venture or who wish to learn about small business management practices. (VS)

153. **State of Small Business: A Report of the President.** Small Business Administration. 1982– . Annual.
Tables, charts, graphs, glossary. SBA 1.1/2:

Contains the text of the President's report on small business and is the most comprehensive source of statistics on small business in the United States. Topics covered include the state of small business, small business financing, veterans in business, and small business as self-employment. Other sections provide statistics on women-owned businesses, minority-owned businesses, the changing characteristics of workers and the size of businesses, procurement, and the dimensions of small business. This is a much more useful reference tool than the *Annual Report* of the SBA (see item no. 146). A glossary of terms is provided, as is a detailed index to assist in finding statistics on specific small-business-related topics. (VS)
Indexed in: *ASI*

154. **Tax Guide for Small Business.** Internal Revenue Service. 1956– . Annual.
T 22.19/2:Sm1

Detailed set of tax instructions for small businesses. Eight sections cover tax schemes for types of business organizations, how to figure business assets, gross profit, net income or loss, disposal of business assets, what constitutes business activity, and how to calculate credits and other taxes, in addition to providing completed sample tax forms. An important reference tool for any small businessperson, although other tax materials from commercial publishers, as well as the IRS, should also be consulted. (VS)

X. PATENTS AND TRADEMARKS

155. **Attorneys and Agents Registered to Practice Before the U.S. Patent and Trademark Office.** Department of Commerce. Patent and Trademark Office. Irregular. Title varies.
C 21.9/2:

Part one of this 2-part directory is an alphabetical listing by name of attorneys and agents registered to practice before the Patent and Trademark Office. Part two is a geographical list of attorneys and agents. Mailing addresses are given in both sections. (MM)

156. **Official Gazette of the U.S. Patent Office: Patents.** Department of Commerce. Patent and Trademark Office. 1872– . Weekly.
Illustrations. C 21.5:

Provides abstracts for all patents granted, lists of all patents reissued, and published notices from the Patent Office. Contained trademarks registered during the week until volume 883, 1971, when the Trademarks began a separate *Official Gazette of the U.S. Patent Office: Trademarks* (see item no. 157). Abstracts generally include at least one illustration. Contents page in each issue explains how to acquire full copies of patents. Weekly indexes and annual cumulative indexes are by list of patentees or by classification of patents. To use the Classification of Patents Index, a subject classification scheme assigned by the Patent Office, it is necessary to use the *Index to the U.S. Patent Classification* (C 21.12/1:) and the *Manual of Classification* (C 21.12:). Indexing is also provided by the Patent Office on CASSIS, the online database available at patent

depository libraries. The *Annual Indexes* are sold separately (C 21.5/2:). (MM)
Indexed in: CASSIS, Self

157. **Official Gazette of the U.S. Patent Office: Trademarks.** Department of Commerce. Patent and Trademark Office. 1971– . Weekly.
Illustrations. C 21.5/4:

Lists Patent and Trademark notices, trademarks published for opposition, and registrations issued, renewed, or cancelled. Marks Published for Opposition are trademarks applied for and include the mark or illustration for each, arranged by product classification numbers. The annual index is by name of registrant. Formerly included with the *Official Gazette of the U.S. Patent Office: Patents* from 1871 to 1970 (see item no. 156). (MM)
Indexed in: Self

XI. GOVERNMENT GRANTS AND CONTRACTS

158. **Catalog of Federal Domestic Assistance.** Executive Office of the President. Office of Management and Budget. 1971-   . Annual (subscription service with updates).
PrEx 2.20: F 31

The catalog provides information on federal programs and federal financial assistance. It is used to identify assistance programs and to obtain general information about programs including agency responsibility, objectives, eligibility requirements, and application and award processes. Indexing is by applicant eligibility (individual, state, etc.), function, and subject. (MM)
Indexed in: Self

159. **Commerce Business Daily.** Department of Commerce. Industry and Trade Administration. 1954-   .
C 57.20:

Provides a list of U.S. government procurements sought, sales and contract awards, surplus property sales, and research and development sources sought. This is a valuable source for businesses planning to bid on U.S. government purchases or contracts, and is a useful reference tool in libraries serving potential contractors. Names and addresses are provided for contact people. The arrangement is by general categories under services and supplies, equipment and materials. No index is provided, but it is available on Dialog as Files 194 and 195. (MM)

160. **Federal Acquisition Regulations.** Department of Defense. 1984-   . Looseleaf (subscription service).
D 1.6/11: (Also issued as *CFR* Title 48)

Government Grants and Contracts 65

This manual and its supplements contain regulations used by federal agencies requesting supplies and services. It covers contracting regulations which would enable companies to do business with the government. Replaced *Federal Procurement Regulation System, Defense Acquisition Regulation,* and *NASA Procurement Regulation.* (MM)

161. **100 Companies Receiving the Largest Dollar Volume of Prime Contract Awards.** Department of Defense. 1969/70– . Annual.
   D 1.57:

This annual report lists summary data on the 100 companies, including subsidiaries, which were awarded the largest total dollar volume of Department of Defense prime contract awards during the year. (MM)

162. **U.S. Government Purchasing and Sales Directory.** Small Business Administration. Office of Procurement and Technical Assistance. Irregular.
   SBA 1.13/3:

This directory is a major aid for any small business that wants to sell to the federal government. The parts and their contents are: Part I, Information on Selling to the Government and to Government Contractors; Part II, Products and Services Bought by Major Military Purchasing Offices and Where to Contact the Offices; Part III, Local Purchases by Military Installations; Part IV, Products and Services Bought by the Major Federal Civilian Purchasing Offices; Part V, Research and Development Opportunities; Part VI, Specifications and Standards; Part VII, Government Property Sales; and Part VIII, Forms. Published irregularly, the 10th edition was issued in 1984. (MM)

XII. PUBLIC FINANCE

163. **Census of Governments.** Department of Commerce. Bureau of the Census. 1957– . Quinquennial.
Tables, graphs. C 3.145/4:

The 1982 *Census,* consisting of 18 reports, presents statistical data in four broad areas: governmental organization, property taxation, public employment, and government finances. The level of coverage varies with each report, but may include national, state, county, municipalities over 5,000 in population, and various districts (public schools, special districts). Governmental Organization section presents summary data on the various levels of government by size classes and types, as well as a helpful narrative description of the governmental structure of each state. The Property Taxation section covers property values and assessment/sales price ratios for states, counties, and cities with populations of over 50,000. Public employment data provide detailed information on the employment and payrolls of local governments, various public employment sector topics, and labor/management relations in state and local governments. The Governmental Finances portion contains statistics on school districts for systems enrolling more than 5,000 students; special district data for selected large districts; and financial data for county and local governments arranged by size and type. Two reports in the Topical Studies section for 1982 are also of reference interest: one details governmental finances for U.S. territories and the other is devoted to graphic presentations of summary financial data. (RB)

164. **Government Finances (GF Series).** Department of Commerce. Bureau of the Census. 1964– . Annual. *Governmental Finances in the United States*
Tables. C 3.191/2-

This series contains the Annual Survey of Government Finances results collected into eight reports which provide revenue, expenditure, debt, and asset data at the federal, state, and local government levels. Because of their diversity and recent changes in the handling of the series by the Government Printing Office, each report is briefly annotated separately.

GF No.1 *State Government Tax Collections* (/2-8:). Summary tables provide revenue totals by type of tax, allowing state comparisons and selected year-to-year comparisons. Tax collection tables for each state provide sales and license revenue, individual and corporation income taxes, property and miscellaneous taxes detailed beyond the broad headings.

GF No.2 *Finances of Employee-Retirement Systems of State and Local Governments* (/2-2:). Reports only on retirement systems of governmental units whose membership is made up of public employees compensated with public funds. State and local summary data are available for receipts, payments, and membership. More detailed information is on the county and city level for retirement systems with more than 200 members.

GF No.3 *State Government Finances* (/2-3:). Summary figures on government revenue, expenditures, indebtedness, and cash and security holdings. The four areas are detailed further in following tables, e.g., selected per capita state government revenues by states.

GF No.4 *City Government Finances* (/2-5:). National totals for municipal finances are arranged by population-size groups. The emphasis of the report is on individual city financial data for cities with a population of over 50,000. Standard revenue and expenditure items are detailed.

GF No.5 *Government Finances* (/2-4:). Tables reporting revenue, expenditures, indebtedness, and cash holdings are arranged to allow comparison of federal, state, and local government totals. State and local tables provide more detailed figures including summary data for special districts and school districts.

GF No.6 *Local Government Finances in Selected Metropolitan Areas and Large Counties* (/2-9:). Finances of MSAs and their county areas are profiled in tables similar to those of GF No.4. Comparison of local governments within each metropolitan area is possible, including county areas.

68 Public Finance

GF No.8 *County Government Finances* (/2-7:) Detailed finances of counties over 100,000 and over 500,000 in population are reported. National summary data are given for all counties. Of particular interest is a per capita comparison of revenue and expenditures by type for counties over 100,000.

GF No.10 *Finances of Public School Systems* (/2-6:). National summary totals for public school system finances; state totals for revenue by source and expenditures by level of education. Detailed tables profile individual districts with enrollments of over 15,000. Standard financial data are included as well as per pupil amounts for various revenue and expenditures. (RB)

165. **Monthly Statement of Public Debt of the United States.** Department of the Treasury. 1974– . Monthly.
Tables. T 1.5/3:

Summary of public debt outstanding for the current month with comparative figures for the same date one year before. Table gives average interest rate and amount outstanding on interest-bearing and noninterest-bearing debts by type. The major section of the approximately 23-page monthly details interest-bearing loans by title and rate of interest, providing dates of issue and redemption, when payable, amount of interest payable, amount issued and outstanding. Prompt source for information on treasury bills, notes, and bonds, and U.S. Savings Bonds. Included with *Daily Statement of the Treasury* prior to July 1974. (RB)

166. **Significant Features of Fiscal Federalism.** Advisory Commission on Intergovernmental Relations. 1976– . Annual.
Tables. Y 3.Ad9/8:18/

Source of extensive information on federal, state, and local revenues, expenditures, tax rates, tax trends, grants-in-aid, employment and earnings, and other related subjects. The first section features historical tables allowing regional or state-by-state statistical comparisons, as well as narrative comparisons of topics such as state income tax laws. A new section with the 1985/86 edition is dedicated to tables ranking the states on various expenditures and revenue items, e.g., per capita direct higher education expenditures. Further comparison of states is provided by the State Profile section where major revenue sources, expenditures, and distribution percentages are reported for each state for seven selected years. The annual has broad application as a source of public finance data. (RB)
Indexed in: Self

167. **United States Government Annual Report.** Department of the Treasury. Financial Management Services. 1894– .
Combined Statement of Receipts, Expenditures, Balances of Government for Fiscal Year (1940–83)
Annual Reports, Division of Bookkeeping and Warrants (1894–1940)
Tables, charts, graphs. T 1.1/3:

Overview of the receipts and outlays of the U.S. Government; may be considered an "Annual Report in Brief." Financial details supporting this summary appear in the *Annual Report Appendix* (T 1.1/8:). Graphs and tables are useful for a broad presentation of the government's financial position. Budget receipts by source, outlays by function and appropriations, and outlays and balances by major department are summarized, as well as a 5-year comparison of financial activity. Brief financial statements are included. (RB)

XIII. TAXATION

168. **Cumulative List of Organizations Described in Section 170(c) of the Internal Revenue Code of 1954.** Department of Treasury. Internal Revenue Service. 1950/52– . Annual, with three quarterly cumulative supplements. *Cumulative List of Organizations, Contributions to which are Deductible Under Section 23(o) and Section 23(q) of the Internal Revenue Code as Amended and the Corresponding Sections of Prior Revenue Acts* (pre-1954)
T 22.2/11:

A cumulative list of organizations to which contributions are tax deductible. The publication lists the legal name of the organization and its location. There is no separate listing of common or popular names of organizations. The listing for the central organization also lists the subordinate units of that organization to which contributions are deductible. The second part of the publication contains a general explanation of the rules covering income tax deductions for contributions by individuals. The list of exempt organizations is not all-inclusive. Status changes in whether or not an organization is exempt appear in the *Internal Revenue Bulletin* (see item no. 172) between editions of the *Cumulative List.* (DC)

169. **Exempt Organizations Current Developments.** Department of the Treasury. Internal Revenue Service. 1983– . Quarterly. Bibliography. T 22.50:

Contains separate sections with brief annotations on exempt organizations' current developments by type, such as legislation by public law number and by bill number; rules and regulations; revenue rulings, procedures, and actions taken by the IRS on decisions made;

current litigation by name of case; General Counsel memoranda; private letter rulings and technical advice given to queriers by the IRS; and general announcements, notices of articles, and publications of interest. (DC)

170. **Governments Quarterly Report, GT. Quarterly Summary of Federal, State and Local Tax Revenue.** Department of Commerce. Bureau of the Census. 1963– . Quarterly. *Government Quarterly Report, GT. Quarterly Summary of State and Local Tax Revenue* (1963–July/Sept. 1982)
Tables. C 3.145/6:

Data on federal, state, and local tax revenue collected by type of tax, state, and certain counties, and local collections of property taxes for selected local areas. Appendixes include descriptions of changes in state tax laws and administrative procedures. More detailed figures on state tax revenue appear in *State Government Tax Collections* (C 3.191/2-8:) and *State Government Finances* (C 3.191/2-3:) (see item no. 164 for both). State and local tax data reported annually in the Bureau's *Government Finances* (GF) series (see item no. 164). Supplemented by *Governments Quarterly Report* (GT) and *Quarterly Summary of State and Local Tax Revenue* (C 3.145/6:). (DC)

171. **Individual Income Tax Returns.** Department of the Treasury. Internal Revenue Service. Statistics of Income Division. 1916– . Annual.
Statistics of Income. Part 1. Compiled from Individual Income Tax Returns and Gift Tax Returns. (1916–53, with minor variations)
Statistics of Income. Individual Income Tax Returns. (1954–82)
Tables, graphs, facsimiles. T 22.35/8:

Final detailed annual tabulation of individual income tax returns. The report contains data on sources of income, adjusted gross income, number of returns, exemptions, deductions, tax credits, self-employment tax, and tax rates. Classifications are by tax status of individual, size of income, marital status, and form of deduction. Also includes information on high-income tax returns. Beginning with the 1983 results, the scope of this report was changed. It no longer includes any analysis of the data. In previous years, the report provided an in-depth description of tax law changes, but after 1983 the only changes described are those that affect the data. The IRS plans to improve the report for tax year 1986 by developing an annual *Source Book* that will provide detailed data by several demo-

72   Taxation

graphic characteristics such as age, filing status, and type of form filed. Tax law changes that would affect the data published will be included, along with any other published articles on individual tax data for that specific year. Preliminary statistics on individual income tax returns are published in the *SOI Bulletin* (see item no. 176). (DC)

172.   **Internal Revenue Bulletin.** Department of the Treasury. Internal Revenue Service. 1922- . Weekly.
Tables, charts. T 22.23:

Contains announcements of official IRS rulings and Treasury decisions, as well as executive orders, court decisions, and laws which deal with taxation and the IRS. Also announces proposed regulations. Special section lists attorneys and agents disbarred or suspended from practice before the Department of the Treasury. The *Bulletin* is cumulated semiannually in the *Internal Revenue Cumulative Bulletin* (see item no. 173). (DC)

173.   **Internal Revenue Cumulative Bulletin.** Department of the Treasury. Internal Revenue Service. 1922- . Semiannual.
Internal Revenue Bulletin. Cumulative Bulletin. (pre-1969)
Tables, charts. T 22.25:

Cumulates the contents and consolidates all items of a permanent nature published weekly in the *Internal Revenue Bulletin* (see item no. 172). The *Cumulative Bulletin* contains three parts. Part 1 includes rulings and decisions based on provisions of the IRS Code of 1954. Part 2 deals with tax treaties and tax conventions, as well as tax legislation and related committee reports. Part 3 consists of administrative and procedural announcements and miscellaneous notices. Included is a list of agents and attorneys disbarred or suspended from practice before the IRS. All notices of proposed rulemaking as well as final rules are published immediately upon issue in the *Federal Register* (see item no. 7). The *Cumulative Bulletin* indexes itself. It is also indexed and supplemented by the *Bulletin Indexes-Digest Supplement System.* Each of the *Index-Digest* services consists of a basic volume and a cumulative supplement that provide lists and digests of items published in the *Cumulative Bulletin,* plus topical indexes of tax-related public laws, tax conventions, and Treasury decisions. The *Index-Digest* services are: Service 1, Income Tax (T 22.25/5:); Service 2, Estate and Gift Tax (T 22.25/6:); Service 3, Employment Tax (T 22.25/7:); Service 4, Excise Taxes (T 22.25/8:). (DC)

Taxation 73

174. **Monthly Treasury Statement of Receipts and Outlays of the United States Government.** Department of the Treasury. Financial Management Service. 1953–   . Monthly.
Monthly Statement of Receipts and Outlays of the United States Government (1972–74)
Monthly Statement of Receipts and Expenditures of the United States Government (1953–72)
Charts, tables. T 63.113/2:

Report of U.S. Government receipts and outlays based on reports filed with the Treasury Department from government agencies. Each monthly report covers the fiscal year to date. Includes budget receipts by source, and outlays by government agency and program. The September 30 issue is entitled *Final Monthly Treasury Statement,* and covers the entire fiscal year, comparing the data to the previous fiscal year. The *Monthly Treasury Statement* is a companion to the *Daily Statement of the United States Treasury* (T 1.5:), which provides data on the cash and debt operations of the Treasury based on reports from the Federal Reserve Banks. Annual figures on budget receipts and outlays are published in *Combined Statement of Receipts, Expenditures, and Balances of the U.S. Government for Fiscal Year* (T 63.113:). (DC)

175. **Statistics of Income. Corporation Income Tax Returns.** Department of the Treasury. Internal Revenue Service. Statistics of Income Division. 1916–   . Annual.
Statistics of Income, Part 2, Corporation Income Tax Returns (1951–54)
Statistics of Income, Part 2, Compiled from Corporation, Income and Excess Profits Tax Returns and Personal Holding Company Returns (1919–51 with minor variations)
Graphs, tables, charts. T 22.35/5:

Detailed report on both foreign and domestic corporate income. Contains data by industry, classified by an industry code similar to the SIC 2- or 3-digit levels. Data are also classified by size of total assets and size of business receipts. Includes statistics on industries' assets, liabilities, receipts, deductions, net income, income subject to tax, credits, and stockholder distributions. Data are compiled from corporation income tax returns. More detailed statistics for industries are available in *Corporation Income Tax Returns, Source Book* (T 22.35/5-2:). Two new, related publications were developed by the IRS Statistics of Income Division in 1985. The first is *Statistics of Income. Partnership Returns* (T 22.35/6:), a compendium of *SOI*

74   Taxation

Bulletin articles, unpublished tabulations, detailed sample descriptions, and return facsimiles. The first release of this compendium covered 1957–83. A 1983–87 edition is to be released in 1989. The second new publication is *Statistics of Income, Source Book, Partnership Returns* (T 22.35/2:P 25/2/); similar in purpose to the *Corporation Source Book* in that it provides detailed industry information. First edition covers 1957–83. Both partnership products replace the annual *Statistics of Income Partnerships* (T 22.35/6:), which was discontinued in 1980. (DC)

176.   **Statistics of Income. SOI Bulletin.** Department of the Treasury. Internal Revenue Service. 1981–   . Quarterly.
Graphs, tables. T 22.35/4:

Provides articles and statistical tables on individual and business income, assets, and expenses, compiled from federal income tax returns. Since its inception in 1981, the *SOI Bulletin* has become the Service's primary informational publication on income statistics. Many of the old *Preliminary Statistics of Income* serial publications were superseded by this periodical. This is the first place that many of the IRS's income statistics appear. The *SOI Bulletin* was formed by the union of: (1) *Preliminary Statistics of Income, Individual Income Tax Returns;* (2) *Preliminary Statistics of Income, Business Income Tax Returns, Sole Proprietorships, Partnerships;* and (3) *Statistics of Income Preliminary, Corporation Income Tax Returns.* The *SOI Bulletin* contains four to six articles on taxation topics per issue. More comprehensive final data are published in the individual *Statistics of Income* annual reports. (DC)

XIV. CONSUMERS

177. **Annual Report.** Consumer Product Safety Commission. 1975?– . Annual.
 Tables. Y 3.C76/3:1/

A two-part report of the Consumer Product Safety Commission. Part one is a narrative account of the Commission's activities and its accomplishments in increasing consumer safety. It includes brief articles on such subjects as compliance and enforcement, hazard identification and analysis, and current developments in regulation. Part two is an extensive collection of appendixes. Regularly included are Estimates of Injuries from Consumer Products; Policies, Regulations, and Proposed Regulations; CPSC Meetings of Substantial Interest; Log and Status of Petitions; Litigation, Advisory Committees, and an Index of Products Regulated by CPSC. (JN)
Indexed in: *ASI*

178. **Consumer Information Catalog.** General Services Administration. Consumer Information Center. Summer 1977– . Quarterly.
 GS 11.9:

A quarterly publication designed to promote the distribution of federal consumer information booklets. Each quarter a selection of booklets is listed with short descriptive annotations and ordering information. Categories are (1) Careers and Education; (2) Children (including child care); (3) Federal Benefits; (4) Financial Planning; (5) Food; (6) Health; (7) Housing; (8) Money Management; (9) Small Business; (10) Travel and Hobbies; (11) Miscellaneous; and a list of participating agencies. (JN)

76   Consumers

179. **Consumer Prices, Energy and Food.** Department of Labor. Bureau of Labor Statistics. Sept. 1980– . Monthly.
Tables. L 2.38/7:

A monthly statistical report on food, gasoline, and utilities. Comparative information is given for the past two months in tabular form on average prices by selected cities and by region. Utilities include gas, electricity, and fuel oil. Food is presented by categories that are broken down very specifically (e.g., round steak). The front page is a condensation of statistics; the last page includes Technical Notes, which is a commentary on the month's prices. (JN)
Indexed in: *ASI*

180. **FDA Consumer.** Department of Health and Human Services. Food and Drug Administration. 1967– . Monthly. *FDA Papers* (1967–72)
HE 20.4010:

This is the official magazine of the Food and Drug Administration for use by the general public. Included are articles on food or drug topics of interest to the layperson such as obesity, cosmetic allergies, breastfeeding, osteoporosis testing, etc. Regular columns include a collection of items gathered from FDA news releases, reports from FDA investigations, and summaries of court actions which involve seizure or injunction proceedings of foods, drugs, devices, or cosmetics. (JN)
Indexed in: *IUSGP, PAIS, RG*

181. **FDIC Consumer News.** Federal Deposit Insurance Corporation. Division of Bank Supervision. 1981– . Quarterly (irregular).
Y 3.F31/8:24

A newsletter of short articles designed to help the average consumer deal with banks. One article per issue, such as Balancing Check Books or Consumers Should Comparison Shop for Credit Terms, appears in Spanish and English. Regular features include Questions from Bank Customers in a question-and-answer format. Issues often concentrate on a particular subject such as deposit insurance or credit. (JN)

182. **Family Economics Review.** Department of Agriculture. Family Economics Research Group. 1943– . Quarterly. *Rural Family Living* (1943–57)
Charts, graphs, tables. A 77.708:

Each issue of this journal has three to four in-depth articles on family economic issues such as economic outlook for families, child care expenditures, food away from home expenditures, and bulk food costs. Research reports published by various government agencies on family economic subjects are abstracted and interspersed throughout the journal. Regular features include statistical charts on the Cost of Food at Home, Consumer Prices, Updated Estimates of the Cost of Raising a Child, and a list of new USDA publications related to family economics. Issue four includes an index to articles for the year. (JN)

Indexed in: *ASI, IUSGP, PAIS,* Self.

183. **Food News for Consumers.** Department of Agriculture. Food Safety and Inspection Service. Jan. 1980– . Quarterly. A 110.10:

This magazine has brief, informative articles for the layperson. Each issue includes a feature article and several short articles on food safety, health, and nutrition topics, and a consumer information column in question-and-answer format. Current news items on new food processes and USDA announcements related to food safety are included in the News Wire column. (JN)

TITLE INDEX

Numbers refer to item entries.

Advance Data from Vital and Health Statistics 32
Agricultural Outlook 120
Agricultural Prices 121
Agricultural Statistics 122, 131
Agricultural Supply & Demand Estimates 129
Agriculture Yearbook 131
Air Carrier Financial Statistics 95, 98
Air Carrier Traffic Statistics 96, 98
Annual Airline Statistics 98
Annual Economic Report (CEA) 17
Annual Economic Review (CEA) 17
Annual Energy Review 111
Annual Housing Survey: United States and Regions 63
Annual Indexes (DOC) 156
Annual Report (Army Chief of Engineers) 105
Annual Report (CPSC) 177
Annual Report (EIA) 111
Annual Report (EEOC) 134
Annual Report (FCC) 60
Annual Report (ICC) 90
Annual Report (SBA) 146, 153
Annual Report Appendix 167
Annual Report on the Statistics of Railways 94
Annual Reports, Division of Bookkeeping and Warrants 167
Annual Statistical Digest (FRB) 12
Annual Statistical Supplement (BLS) 140
Annual Survey of Manufactures 67, 69
Appendix (U.S. Budget) 2

Area Handbook Series 33
Area Wage Survey 135
Attorneys and Agents Registered to Practice Before the U.S. Patent and Trademark Office 155
Availability of Heavy Fuel Oils by Sulfur Level 79

B.C.D. 13
Background Notes 34
Banking and Monetary Statistics, 1949–70 12
Block Statistics 27
Budget in Brief 1
Budget of the United States Government 2
Bulletin Indexes-Digest Supplement System 173
Bureau of Economic Analysis Staff Papers 3
Bureau of the Census Catalog 4
Business America 35
Business Basics 147
Business Conditions Digest 13, 20
Business Cycles Development 13
Business Statistics 14

CAA Statistical Handbook of Aviation 97
CPI Detailed Report 15
Capacity Utilization Manufacturing and Materials 68
Catalog of Federal Domestic Assistance 158
Catalog of U.S. Census Publications 4
Census Catalog and Guide 4
Census of Agriculture 123

78

Title Index

Census of Business 82, 83, 86
Census of Construction Industries 64
Census of Governments 163
Census of Manufactures 69, 67
Census of Mineral Industries 70
Census of Population 26
Census of Population and Housing 27
Census of Retail Trade 82, 84
Census of Service Industries 86
Census of Transportation 91
Census of Wholesale Trade 83
Census Tracts 27
City Employment in 138
City Government Finances 164
Code of Federal Regulations 5, 7, 11
Combined Statement of Receipts, Expenditures, Balances of Government 167, 174
Commerce America 35
Commerce Business Daily 159
Commerce Today 35
Congressional District Data Book 10
Congressional Districts of the _th Congress 27
Construction Reports 65
Construction Review 66
Consumer Credit Outstanding at Finance Companies 56
Consumer Income 28
Consumer Information Catalog 178
Consumer Price Index 15
Consumer Prices, Energy and Food 179
Corporation Income Tax Returns, Source Book 175
Cost and Quality of Fuels for Electric Utility Plants 114
Cost of Food at Home 180
Country Market Survey 36
County and City Data Book 6, 9, 10
County Business Patterns 52
County Employment in 138
County Government Finances 164
Crop Production 124
Crude Petroleum, Petroleum Products 78
Cumulative List of Organizations Described in Section 170(c) of the Internal 168
Current Business Reports. BR, Monthly Retail 84
Current Business Reports. BW, Monthly Wholesale 85
Current Business Reports. Wholesale Trade Report 85
Current Industrial Reports 53
Current Mortality Analysis 30
Current Population Reports 28

Current Population Reports, P-57 137
Customs Regulations of the United States 37

Daily Statement of the U.S. Treasury 165, 174
Data User News 29
Defense Acquisition Regulation 160
Deliveries of Fuel Oil 78
Descriptive Supplement to Economic Indicators 16
Dictionary of Occupational Titles 136, 143

EPA Cumulative Bibliography 133
EPA Journal 132
EPA Publications Bibliography 133
EPA Reports Bibliography 133
Economic Indicators 16
Economic Indicators of the Farm Sector 125
Economic Report of the President 17
Economic Trends and Their Implications for 39
Electric Energy and Peak Load Data 112
Electric Power Annual 112
Electric Power Monthly 113, 112
Electric Power Quarterly 114, 112
Employment and Earnings 137
Employment and Unemployment in States 145
Energy and Peak Load Data 113
Energy Information Administration Weekly 81
Energy Related Housing Characteristics 63
Estimates of Labor Force, Employment 137
Excess Profits Tax Returns and Personal 175
Exempt Organizations Current Developments 169
Export Briefs 38

FAA Statistical Handbook of Aviation 97
FDA Consumer 180
FDA Papers 180
FDIC Consumer News 181
Family Economics Review 182
Farm Numbers 124
Farm Population 28
Farm Product Prices 121
Federal Acquisition Regulations 160
Federal Procurement Regulation System 160
Federal Register 7, 5, 11, 173
Federal Reserve Bulletin 18, 12
Federal Reserve Chart Book 19, 21

80 Title Index

Federal-State Cooperative Programs for Population Estimates 28
Finance Companies 56
Finances of Employee-Retirement Systems of State and Local Governments 164
Finances of Public School Systems 164
Financial and Operating Statistics: Class I Railroads 106
Financial Characteristics of the Housing Inventory 63
Financial Statistics of Selected Electric Utilities 115
Fisheries of the United States 73, 74
Fishery Industries of the United States 74
Fishery Statistics of the United States 74, 73
Food News for Consumers 183
Forecasts of Citizen Departures to Europe for 109
Foreign Agricultural Trade of the United States 126
Foreign Agriculture 127
Foreign Crops and Markets 127
Foreign Economic Trends 39
Franchise Company Data for Equal Opportunity 148
Franchise Opportunities Handbook 148
Franchising in the Economy 149
Freight Commodity Statistics of Class I Railroads 107

General Housing Characteristics 63
Government Employment 138
Government Finances 164
Governmental Finances 164
Governments Quarterly Report 170

Handbook of Airline Statistics 98
Handbook of Cyclical Indicators 20
Handbook of Labor Statistics 139
Health Care Financing Review 87
Health, United States 88
Highlights of U.S. Export and Import Trade 40
Highway Statistics 100
Historical Chart Book 21, 19
Historical Statistics of the United States 10
Historical Tables 2
Housing Characteristics of Recent Movers 63
Housing Completions 65
Housing Starts 65
Housing Units Authorized by Building Permits 65

Index to the U.S. Patent Classification 156
Indicators of Housing and Neighborhood Quality 63
Individual Income Tax Returns 171
Internal Revenue Bulletin 172, 173, 168
Internal Revenue Cumulative Bulletin 173, 172
International Commerce 35
International Finance 41
Inventory of Power Plants in the United States 116

Key Officers of Foreign Service Posts 42

Local Area Personal Income 22
Local Government Employment in 138
Local Government Finances in Selected 164

MLR 140
Main Line Sales of Natural Gas 76
Major Matters Before the FCC 61
Management of the United States Government 2
Manual of Classification 156
Maritime Subsidies 102
Marketing Research Reports 128
Merchant Fleets of the World 103
Merchant Ships Built in the United States 104
Mineral Resources of the United States 72
Minerals and Materials 71
Minerals Yearbook 72
Monthly Chart Book 19, 21
Monthly Comment on Transportation Statistics 93
Monthly Energy Review 117, 111
Monthly Gasoline Reported by States 75
Monthly Labor Review 140, 137
Monthly Marriage Report 30
Monthly Motor Fuel Reported by States 75
Monthly Petroleum Product Price Report 77
Monthly Petroleum Statement 79
Monthly Petroleum Statistics Report 79
Monthly Product Announcements 4
Monthly Report of Air Carrier Traffic Statistics 96
Monthly Report of the U.S. Bureau of Labor Statistics 137
Monthly Report on the Labor Force 137
Monthly Review of the U.S. Bureau of Labor Statistics 140
Monthly Statement of Public Debt 165

Title Index   81

Monthly Supplement of the Commerce Reports 25
Monthly Treasury Statement of Receipts and Outlays 174
Monthly Vital Statistics Bulletin 30
Monthly Vital Statistics Report 30, 32
Motive Power and Car Equipment of Class I Railroads 106

NASA Procurement Regulation 160
National Income and Product Accounts of the United States 23
National Transportation Statistics 92
National Trends in Electric Energy Resources 113
Natural and Synthetic Gas 76
Natural Gas Annual 76
Natural Gas Monthly 76
New One-Family Houses Sold and for Sale 65
New Residential Construction in SMSAs 65
New Ship Construction 104

Occupational Outlook Handbook 141
Occupational Outlook Quarterly 141
Official Gazette of the U.S. Patent Office: Patents 156, 157
Official Gazette of the U.S. Patent Office: Trademarks 157, 156
100 Companies Receiving the Largest Dollar Volume of Prime Contract Awards 161
Operating Revenues and Operating Expenses of Class I Railroads 106
Operating Statistics of Large Railroads 106
Overseas Business Reports 43

Personnel Literature 142
Petroleum Marketing Monthly 77
Petroleum Refineries in the U.S. 78
Petroleum Supply Annual 78, 79
Petroleum Supply Monthly 79, 78
Population Characteristics 28
Population Estimates and Projections 28
Post Reports 44
Power Production, Fuel Consumption, and 112
Preliminary Power Production 113
Preliminary Statistics of Income 176
Price Index of New One-Family Houses Sold 65
Prices and Margins of No. 2 Distillate 77
Public Employment in 138
Public Health Reports 89

Public Roads 101

Quarterly Abstract Bulletin 133
Quarterly Financial Report for Manufacturing 54
Quarterly Operating Data of Telegraph Carriers 62
Quarterly Operating Data of Telephone Carriers 62
Quarterly Summary of State and Local Tax Revenue 170

Rail Carload Cost Scale by Territory 108
Receivables Outstanding at Finance Companies 56
Report of the Secretary of Agriculture 131
Residential Alterations and Repairs 65
Residential Electric Bills in Major Cities 113
Revenue Traffic Statistics of Class I Railroads 106
Rural Family Living 182

SEC Monthly Statistical Review 57
SOI Bulletin 171
Sales Finance Companies 56
Sales of Liquefied Petroleum 78
Savings and Home Financing Sourcebook 58
Selected Income and Balance Sheet Items of Class I Railroads 106
Short-Term Energy Outlook 118
Significant Features of Fiscal Federalism 166
Small-Area Data Notes 29
Small Business Management Series 150
Small Business Subcontracting Directory 151
Social Security Bulletin 24
Special Analysis of the Budget 2
Special Censuses 28
Special Studies 28
Standard Industrial Classification Manual 8
Standard Occupational Classification Manual 143
Starting and Managing Series 152
State and Metropolitan Area Data Book 9, 6, 10
State, County, and Selected City Employment 145
State Export Series 45
State Government Finances 164, 170
State Government Tax Collections 164, 170
State of Small Business 153, 146

82   Title Index

Statistical Abstract of the United States 10, 6, 9
Statistical Appendix to Minerals Yearbook 72
Statistical Bulletin (SEC) 57
Statistical Handbook of Civil Aviation 97
Statistical Summary (FHLB) 58
Statistical Supplement to the Survey of Current Business 14
Statistics of Communication Common Carriers 62
Statistics of Electric Utilities in the U.S. 115
Statistics of Income. Corporation Income Tax 175
Statistics of Income. Individual Income Tax 171
Statistics of Income. Part 1. Compiled from Individual 171
Statistics of Income. Part 2. Compiled from Corporation 175
Statistics of Income. Partnership(s) 175
Statistics of Income Preliminary 176
Statistics of Income. SOI Bulletin 176, 171
Statistics of Income. Source Book 175
Statistics of Privately-Owned Electric Utilities 115
Statistics of Publicly-Owned Electric Utilities 115
Statistics of the Communications Industry 62
Summary and Analysis of International Travel to the United States 110
Summary Characteristics for Governmental Units and Standard Metropolitan Statistical Areas 27
Supplement to Economic Indicators 16
Supply, Disposition, and Stocks of All Oils 79
Survey of Current Business 25, 23, 14

TOP Bulletin 47
Tariff Schedules of the United States 46
Tax Guide for Small Business 154
Trade and Employment 144
Train and Yard Service of Class I Railroads 106
Transport Economics 93, 94
Transport Statistics in the United States 94, 93
Treasury Bulletin 59
Typical Electric Bills 119, 113
Typical Residential Electric Bills 119

U.S. Crude Oil, Natural Gas, and Natural Gas Liquid Reserves 80

U.S. Decennial Life Tables 31
U.S. Department of State Indexes of Living Costs Abroad 48
U.S. Foreign Agricultural Trade 126
U.S. Foreign Trade Highlights 49, 40
U.S. Foreign Trade: Highlights of Exports and Imports 40
U.S. Government Manual 11
U.S. Government Purchasing and Sales Directory 162
U.S. Import and Export Price Indexes 50
U.S. Imports and Exports in the United States 76
U.S. Imports and Exports of Natural Gas 76
U.S. Industrial Outlook 55
U.S. International Air Travel Statistics 99
U.S. Life Tables 31
U.S.D.A. Statistical Bulletin 125
U.S.D.A. Yearbook 131
Underground Natural Gas Storage in the U.S. 76
Unemployment in States and Local Areas 145
United States Department of Agriculture Yearbook 131
United States Government Annual Report 167
United States Government Manual 11, 7
United States Government Organization Manual 11
United States Trade Performance in 51
Urban and Rural Housing Characteristics 63

Value in New Construction Put in Place 65
Vital Statistics of the United States 30, 32

Waterborne Commerce of the United States 105
Weekly Petroleum Status Report 81
World Agricultural Situation 130
World Agricultural Supply and Demand Estimates 129
World Agriculture Outlook and Situation 130
World Agriculture Situation and Outlook Report 130

Yearbook of Agriculture 131, 122
Yearbook of the United States Department of Agriculture 131

SUBJECT INDEX

Numbers refer to item entries. Users seeking information on industries should look under type of industry (i.e., manufacturing) as well as specific industry name.

Administrative agencies 11
Aeronautics 97
Agricultural forecasting 120
Agriculture 17, 120, 131
 finance 125
 international aspects 127, 129, 130
 statistics 122, 123
Air travel 99
Airlines 95, 97
 finance 95
 statistics 96, 98
Area studies 33, 34, 44
Automobile loans 56

Balance of payments 25, 41
Bank reserves 19
Bankruptcy 146
Banks and banking 12, 18, 19, 21
Birth rate 30, 32
Broadcasting industry 60, 62
Budget, U.S., 1, 2, 167, 174
Building contracts 66
Building permits 66
Bus lines 90, 91, 94
Business cycles 13
Business failures 146

Cable television 60
Capital investments 13, 16, 25
Cargo preference 102
Cash management 57
Charts 13, 19, 21
Chemical industry 14, 53

Child rearing, costs 182
Clothing industry 53
Coal 111, 117
Coal industry 14, 55
Coatings 55
Collecting of accounts 59
Commodities, prices 14, 25
Communication industries 14
 laws and regulations 60, 61
Comparative economics 33, 39
Computer industry 55
Construction industry 14, 16, 19, 55, 64, 65, 66
Consumer education 178, 180, 181
Consumer price indexes 15
 international aspects 48
Consumption (economics) 13, 15, 16
Container industry 55
Contracts, government 161
 subcontracting 151
Corporate income tax 175, 176
Corporate profit 12, 13, 16, 17, 25
Corporations, finance 18
Counties 52
Credit 12, 13, 17, 18
Crops, production 124
Customs administration 37

Debt 21
Debts, external 41
Debts, public 1, 2, 165
Diplomatic and consular service 42, 44
Discrimination in employment 134

83

84 Subject Index

Divorce 30, 32
Drug industry 55
Drugs 180

Economic growth 21
Economic indicators 12, 16, 20
  international aspects 39
Economic research 3
Education, finance 164
Electric industries 55, 112, 113, 114, 115
  finance 115
Electric power 111, 117
Electric power plants 114, 116
Electric rates 113, 119, 179
Employment 13, 14, 16, 17, 23, 25, 52, 137, 139, 140, 144, 145
  international aspects 139
Energy consumption 92
Energy resources 111, 118
Environmental protection 132
Exhibitions and fairs 47
Expenditures, public 1, 2, 174
Export-import trade 14, 35, 43, 144
  laws and regulations 43
  promotion 35, 36, 43, 47
  statistics 40, 45, 49, 51

Family 182
Farm produce 124
  export-import trade 38, 126, 127
  marketing 128
  prices 121
Federal reserve banks 18
Federal Reserve System (U.S.), Federal Open Market Committee 18
Finance 14, 17, 59
Finance companies 56
Fish industry 73, 74
Fisheries 73, 74
Flow of funds 12, 18
Food 180, 183
  prices 179
Food industry 14, 53
Foreign exchange rates 18, 48
Foreign licensing agreements 43
Forest products industry 55
Forwarding companies 90, 94
Franchise system 55, 148
  statistics 149
Furniture industry 53

Gas industry 55
Gasoline 81
  prices 179
Geothermal energy 111
Glass industry 14, 53
Government bonds 12, 165

Government officials and employees 1, 2, 11, 138, 163
Government publications 4
Grants-in-aid 158
Gross national product 1, 2, 16, 21, 23, 25

Health care industry 87, 88
Hours of labor 137
Households 28
Housing 9, 27, 63, 65
  costs 65
  finance 57
Housing starts 65

Income 12, 13, 17, 21, 22, 23, 38, 176
Income tax 171, 176
Industrial capacity 16, 68
Industrial relations 140
Industries, size of 52
Industry 53, 55
  classification 8
  finance 175
Industry and state 5, 7
Infants, mortality 32
Input-output analysis 25
Installment plan, statistics 12, 18
Insurance, unemployment 24
Interest rates 12, 16, 18, 19, 21
  international aspects 19
International finance 41, 59
International Monetary Fund 41
Inventories 16, 25
Investments, foreign 25, 43

Job descriptions 136

Labor 139, 149
Labor productivity 13, 17, 140
Labor supply 21, 139
Leasing and renting of equipment 55
Leather industry 14, 53
Livestock, prices 121
Local finance 164, 166, 170
Local officials and employees 138
Lumber industry 53

Machinery industry 53, 55
Manufacturing industries 6, 9, 54, 55, 67, 68, 69
  statistics 67, 68, 69
Marketing, international aspects 36, 43
Marriage 30, 32
Medical instruments and apparatus 55
Metal products 53
Metals 14
Mining industry 54, 55
Minority business enterprises 153

Subject Index 85

Money 13, 16
   velocity of circulation 19
Money supply 12, 17, 18, 19, 21
Mortality 30, 32
   tables 31
Mortgages 12, 19
Municipal finance 164, 166
Municipal governments 163
Municipal officials and employees 138
   salaries, pensions, etc. 164

National income 13, 16, 17, 23, 25
Nations 34
   politics and government 33, 34, 44
Natural gas 76, 111, 117
   prices 179
Neighborhoods 63
New business enterprises 146
Nonprofit institutions, taxation 169
Nuclear energy 111, 117
   international aspects 117

Occupational health and safety 140
Occupations 141
   classification 136, 143
Oil and gas reserves 80
Oil fuel 81
   prices 179
Options (contracts) 58
Orders 13, 16

Paper industry 14, 53, 55
Patents 156
Payrolls 52, 137
Personal finance 19, 21
Personnel management 142
Petroleum 111, 117
Petroleum industry 14, 55, 77, 78, 79, 81
   export-import trade 79, 81, 111, 117
Petroleum pipelines 94
Petroleum refineries 78, 79, 81
Petroleum supply and demand 78, 79
Plastics industry 53, 55
Pollution control in industry 25
Population forecasting 28
Population statistics 10, 26, 27, 28
   cities 6, 9, 26
   counties 6, 26
   metropolitan statistical areas 6, 9
   states 9, 26
Price indexes 13, 20, 23, 50; *see also*
   Consumer price indexes
Prices 13, 16, 21, 139, 140
Product safety 177
Production 13, 16, 17, 21, 68
Property tax 163, 170

Public finance 6, 19, 163, 164, 166, 170, 174
Public health 89
Public services 63
Public utilities 14
Purchasing, government 159, 160, 162

Radio broadcasting 62
Railroads 90, 94
   costs 108
   finance 106
   freight 107
Rankings 6
Ratio analysis 54
Real property 14, 18
Refuse as fuel 111
Regulation of industry 5, 7
Retail trade 6, 9, 82, 84
Roads 100, 101
Rubber industry 14, 53, 55
Rural population 28

Savings 12, 23, 57
Securities, registration 58
Service industries 6, 9, 86
Shipbuilding 104
Shipping 102
   United States 105
Shipping lines 90, 94
Ships, nationality 103
Small business 146, 152, 153
   management 147, 150, 153
   taxation 154
Social security 24
Social systems 33, 44
Solar energy 111
State finance 164, 166, 170
State governments 163
State officials and employees 138
   salaries, pensions, etc. 164
Stock market 18, 19, 21, 58

Tariff 46
Taxation
   charitable deductions 168, 169
   decisions 172, 173
Telegraph companies 62
Telephone 55, 62
Telephone equipment industry 55
Television broadcasting 60
Textile industry 14, 53
Trade missions, American 47
Trade unions 139
Trademarks 157
Trading companies 54
Transportation 14, 25, 91, 92, 93
Travel
   Europe 109

Subject Index

United States 110
Treasury bills and notes 165
Trolley buses 94
Truck freight service 90, 91

Unemployment 137, 139, 145
United States
  economic conditions 17
  statistics 10
United States. Bureau of the Census 29
United States. Consumer Product Safety Commission 177
United States. Dept. of State. Foreign Service 42, 44
United States. Dept. of the Treasury 174
United States. Environmental Protection Agency 133
United States. Equal Employment Opportunity Commission 134
United States. Federal Aviation Administration 97
United States. Federal Communications Commission 60
United States. Food and Drug Administration 181
United States. Internal Revenue Service 172, 173
United States. Interstate Commerce Commission 90
United States. Patent and Trademark Office 155
United States. Small Business Administration 146

Vital statistics 30
Vocational guidance 141

Wage surveys 135
Wages and salaries 13, 14, 16, 17, 139
Wholesale trade 6, 9, 83, 85
Women entrepreneurs 153
Wood as fuel 111
World Bank 41